# INTERNETWORKING AND COMPUTING OVER SATELLITE NETWORKS

# INTERNETWORKING AND COMPUTING OVER SATELLITE NETWORKS

Edited by
YONGGUANG ZHANG
HRL Laboratories, LLC

**Kluwer Academic Publishers**
Boston/Dordrecht/London

**Distributors for North, Central and South America:**
Kluwer Academic Publishers
101 Philip Drive
Assinippi Park
Norwell, Massachusetts 02061 USA
Telephone (781) 871-6600
Fax (781) 871-6528
E-Mail: < kluwer@wkap.com>

**Distributors for all other countries:**
Kluwer Academic Publishers Group
Post Office Box 322
3300 AH Dordrecht, THE NETHERLANDS
Telephone 31 78 6576 000
Fax 31 78 6576 254
E-Mail: < services@wkap.nl>

Electronic Services < http://www.wkap.nl>

**Library of Congress Cataloging-in-Publication Data**

Internetworking and Computing over Satellite Networks
Yongguang Zhang (Ed.)
ISBN 1-4020-7424-7



*Printed on acid-free paper.*

Printed in the United States of America

# Contents

# List of Figures

# List of Tables

# Preface

Satellite networks will play an increasingly important role in our future information-based society. This trend is evidenced by the large number of systems in operation and in planing, such as DirecPC/DirecWay, Iridium, Spaceway, and Teledesic. The benefits of satellite communications include high bandwidth, global coverage, and untethered connectivity; the services are often real-time, multicast, mobile and rapidly deployable. Services based on satellite communications include telemedicine, public information services, education, entertainment, information dissemination, Internet access, digital battlefield, emergency and disaster response, etc.

Consequently, satellite communications introduce a new set of technical problems in mobile networks and applications. In essence, satellite links have fundamentally different properties than terrestrial wired or wireless networks. These include larger latency, bursty error characteristics, asymmetric capability, and unconventional network architecture. These difference have far-reaching effects on many internetworking and distributed computing issues.

In this collection, we present ten chapters written by active researchers in this field. Some chapters survey the recent work in a particular topic and describe the state-of-the-art technologies; others present the latest research results in a particular technical problem. The order of the chapters follows the ISO network layer model. First, chapter 1 serves as an introduction to the satellite networks and gives an overall picture of its role in our lifes in the information age. Chapter 2 and 3 focus on the network architecture and medium access controls (Layer 2). Chapter 4 and 5 focus on the routing issues related to satellite networks (Layer 3). Chapter 6, 7, and 8 explain TCP and the transport protocol issues (Layer 4). Finally, chapter 9 and 10 study the application issues in data broadcast and information dissemination.

Specifically, chapter 2 introduces a multi-satellite network called satellite constellations. It describes the effects of orbital geometry on network topology and the resulting effects of path delay and handover on network traffic. The design of the resulting satellite network as an autonomous system is also discussed here.

Chapter 3 surveys the medium access control (MAC) protocols for satellite networks. Many such protocols have been designed to handle different types of traffic and meet different performance requirements. This chapter gives a comprehensive comparison of these protocols.

Chapter 4 describes an application of satellite network to deliver terrestrial multicast traffic. It explains how to configure a satellite network to support IP multicast, how to bridge Internet-based multicast sessions to a satellite network. The chapter also gives an analysis of the performance impacts.

Chapter 5 studies a technical problem introduced by satellite networks – unidirectional link routing. The chapter explains the technical challenges of this problem and a practical solution adopted by engineers working in this field.

Chapter 6 moves up to the transport layer and surveys TCP-over-satellite work. It describes the challenges that the satellite network environment poses to TCP performance, and summarizes a number of standard TCP options as well as research proposals that can improve TCP-over-satellite performance.

Chapter 7 focuses on one such technique for improving TCP performance: TCP Performance Enhancement Proxy. This chapter explains how it has become the satellite industry's best practice and why it is still considered controversial among the Internet community.

To better understand this technique, Chapter 8 presents a performance study on TCP Performance Enhancement Proxy. It includes results from both model-based and a measurement-based studies. The chapter also presents the implications of these findings on system design, deployment, and provisioning.

Chapter 9 studies an application of satellite network called data broadcasting and focuses on a important technical challenge: how to determine the broadcast schedule so that the clients receive the best quality of service. This chapter presents a theoretical analysis on the optimal broadcast scheduling problem, and derives a heuristic algorithm for producing near-optimal on-line schedules.

Finally, Chapter 10 describes a satellite-based information dissemination application and addresses another technical challenge: the mismatches in characteristics between satellite and terrestrial networks. The chapter proposes a new model called Intelligent Information Dissemination Service to solve this problem.

The book can be used by students, researchers, and engineers in satellite-related data communication networks. It can also be served as a reference book for graduate students in advanced computer networks and distributed systems study.

Although there are many books on the subject of satellite communications, few covers the data networking and computing aspect in satellite networks. We believe this book can help filling the void with a focus on internetworking and distributed computing issues. Since it is impossible to cover every aspects and all activities in this emerging subject in just one book, I hope it does serve as a sampling on the current state of research and technology development. I hope that you enjoy them.

YONGGUANG ZHANG

# Contributing Authors

**Kevin C. Almeroth** is an Associate Professor and Vice Chair of Department of Computer Science at the University of California in Santa Barbara. His e-mail address is almeroth@cs.ucsb.edu.

**Son K. Dao** is a Research Program Manager and a Chief Technologist at HRL Laboratories, LLC. He is also the Chief Scientific Officer of X-Laboratories and a visiting professor at California State University at Northridge. His e-mail address is skdao@hrl.com.

**Vikram Gupta** is a Masters Student in the Department of Electrical Engineering at the University of California, Riverside.

**Thomas R. Henderson** is a staff researcher at Boeing Phantom Works. He received a Ph.D. from the University of California, Berkeley. His e-mail address is thomas.r.henderson@boeing.com.

**Shu Jiang** is a Ph.D. Student in the Department of Computer Science at Texas A&M University. His e-mail address is jiangs@cs.tamu.edu.

**Srikanth V. Krishnamurthy** is an Assistant Professor of Computer Science and Engineering at the University of California, Riverside. His e-mail address is krish@cs.ucr.edu.

**Chen Liu** is a Graduate Student in the Department of Electrical Engineering at the University of California, Riverside.

**Mingyan Liu** is an Assistant Professor of Electrical Engineering and Computer Science at the University of Michigan, Ann Arbor. Her e-mail address is mingyan@eecs.umich.edu.

**Eddie C. Shek** is the Chief Technology Officer of Vizional Technologies, Inc. He received a Ph.D. from the University of California, Los Angeles. His e-mail address is eshek@vizional.com.

**Nitin Vaidya** is an Associate Professor of Electrical and Computer Engineering at the University of Illinois at Urbana-Champaign. His e-mail address is nhv@uiuc.edu.

**Darrel J. Van Buer** is a Research Scientist in HRL Laboratories, LLC. He received a Ph.D. from the University of California, Los Angeles.

**Lloyd Wood** completed his PhD in internetworking with satellite constellations at the Centre for Communication Systems Research, part of the University of Surrey, while working for Cisco Systems. His e-mail address is L.Wood@eim.surrey.ac.uk.

**Yongguang Zhang** is a Senior Research Scientist in HRL Laboratories, LLC and an Adjunct Assistant Professor of Computer Sciences at the University of Texas at Austin. He received a Ph.D. from Purdue University. His e-mail address is ygz@hrl.com.

Chapter 1

# THE ROLE OF SATELLITE NETWORKS IN THE $21^{ST}$ CENTURY

Son K. Dao
*HRL Laboratories, LLC*

Abstract: In the global information infrastructure (GII), satellite networks will play an increasingly important role because of their unique benefits. This chapter briefly introduces the architecture, the vision, and the challenge of future Internet-over- satellite services and applications.

## 1. INTRODUCTION

Most current Internet backbones and local networks are wired terrestrial networks (e.g., fiber optics, cable, and telephone lines) with emerging terrestrial wireless access (e.g., 802.11 wireless LAN). For the past 10 years, researchers have been working on the next generation Internet that can support high bandwidth applications and ubiquitous computing with mobile/wireless networks. Among these mobile/wireless networks, satellite networks offer great potential for multimedia applications with their ability to broadcast and multicast large amounts of data over a very large area and to achieve global connectivity. Recent advancement and deployment of commercial products in satellite communication networks demonstrate the promise of such ubiquitous access to Internet.

By definition, a satellite network is a data communication network facilitated by one or more earth-orbiting communication satellite(s). It can be divided into two segments: the space segment and the ground segment. The space segment consists of the satellite hardware and the communication payload. The onboard communication equipments are for transmitting and receiving signals to and from the ground. If the satellite network has more than one satellite, the space segment can also include inter-satellite communication links (ISL). The ground segment consists of ground stations and network operation centers (NOC). A ground station has a satellite antenna (usually in the shape of a dish) and other communication hardware and software for transmitting to and receiving from the satellites. It can serve as a network interface to a host computer and enables it to communicate with others through the satellites. The ground stations are also the external interface of the satellite networks when they are part of network routers and gateways connected to other networks. The NOC controls the satellite operation and manages the network resource. A satellite network is a very complex data communication network and involves significant upfront investment and management cost. *Figure 1-1* illustrates the structure of a satellite network.

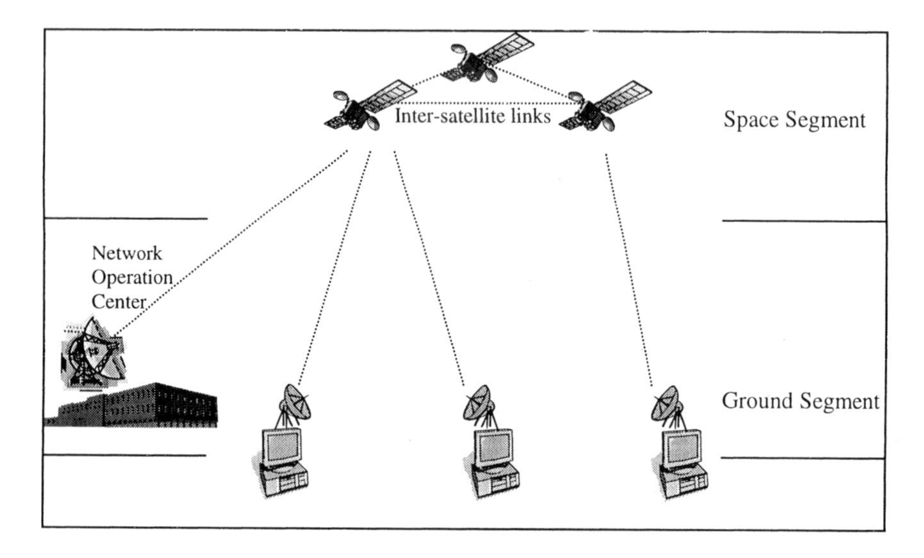


*Figure 1-1.* A satellite network as a data communication network

By the type of satellites and their orbital positions, satellite networks can be categorized into GEO-based, LEO-based, MEO-based, or hybrid. A GEO (geostationary-earth-orbit) satellite positions at 36,000 km above the earth equator and it stays "stationary" relative to surface of the earth (*Figure 1-2*). A LEO (low-earth-orbit) satellite orbits around the world at an altitude of several hundred kilometers to a few thousand kilometers. There are also MEO (medium-earth-orbit) satellites that have an orbit in between of GEO and LEO. Since neither LEO nor MEO satellites can stay at a fixed position relative to surface of the earth, a LEO- or MEO-based satellite network often requires a constellation of multiple satellites to provide uninterrupted service. GEO satellites have the advantage of large footprint (area of coverage) at approximately 1/3 of the world surface, but the latency is much higher, at approximately 250ms. On the other hand, LEO- and MEO-based networks have the advantage of having lower latency, but they face a much harder network management challenge since hand-off, tracking, and routing must be done properly. There are also hybrid satellite networks, where different types of satellites complement each other as one system.

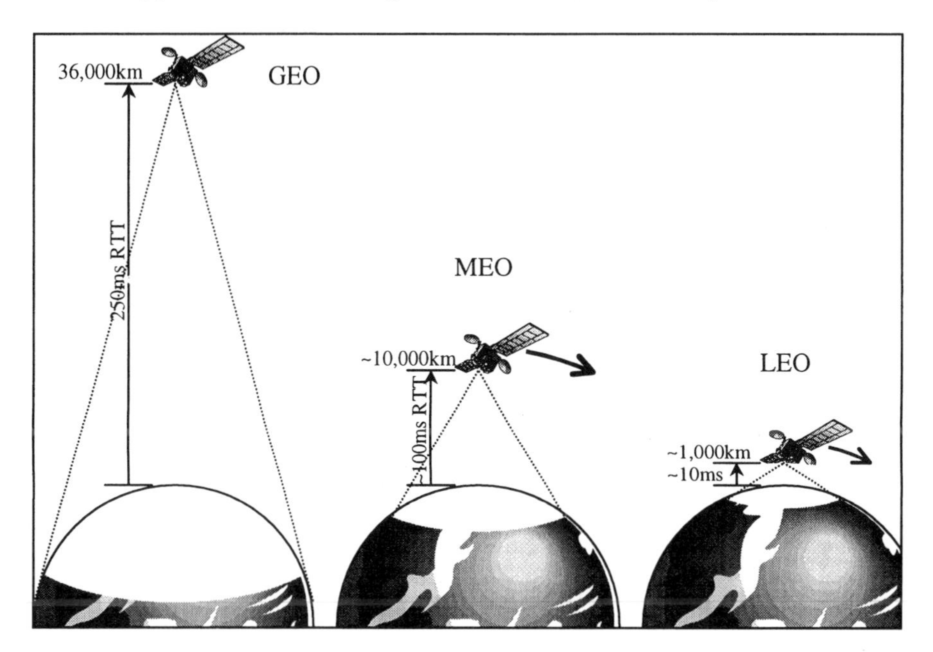


*Figure 1-2.* GEO, MEO, and LEO satellites

## 2. INTERNET OVER SATELLITE ARCHITECTURE

When interconnected with the global Internet, a satellite network can be used to carry Internet traffic and provide Internet services. Internet-over-satellite has the following merits:

- *Ubiquitous coverage.* Theoretically, 3 GEO satellites can coverage the entire world. MEO- or LEO-based networks are intrinsically global networks because the satellites circle around the earth. This complements the terrestrial services, which will not reach or meet the QoS demand in each and every location.
- *Untethered communication.* Satellite communication is inherently wireless. Users can enjoy flexible fixed as well as mobile communications anywhere to anywhere within the footprint (area of coverage) of the satellite network.

- *High bandwidth.* With high-power transponders, wide frequency band, and spot beams, many newer broadband satellite systems are designed to deliver tens of gigabits per second total throughput.
- *Broadcast/multicast capability.* Satellite networks are attractive for broadcast/multicast, or point-to-multi-point applications. The performance is uniform and predictable. In contrast, multicast in the mesh terrestrial network requires complicated multicast routing, where the performance can vary for each multicast group member and is dependent on the route from the source.

The Internet-over-satellite architecture varies by the role of the satellite network in the global Internet, and by the role of the satellite(s) in the satellite network.

## 2.1 The Roles of Satellite Network in the Internet

The satellite network can play different roles in the Internet: as a link technology, as a subnetwork technology, or as a core network (*Figure 1-3*).

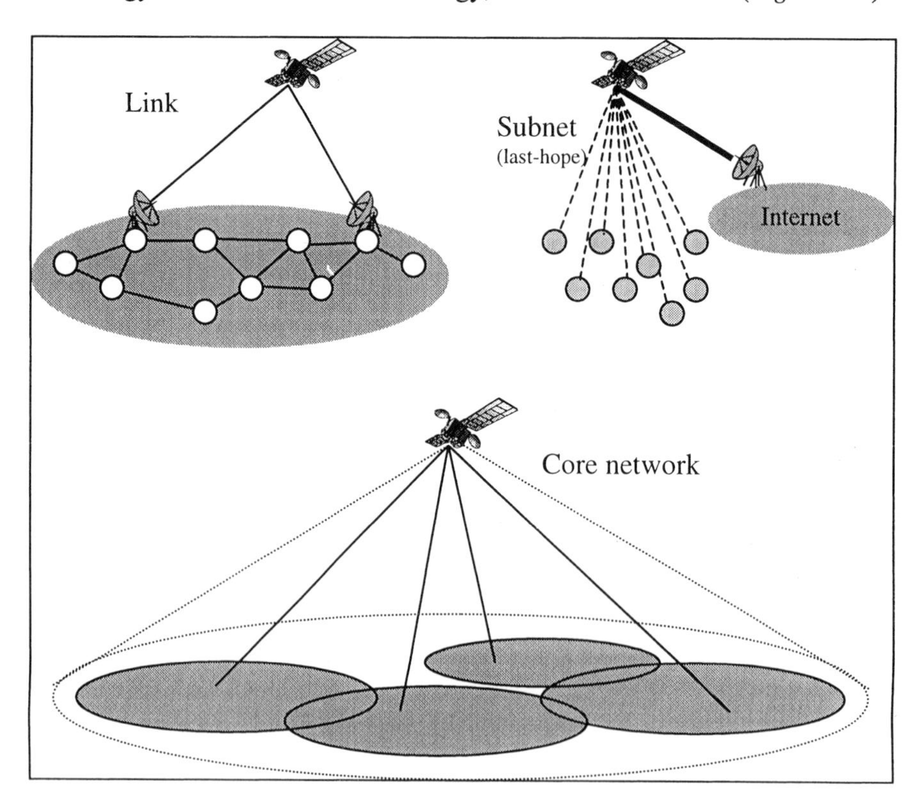


*Figure 1-3.* Satellite network roles in the Internet

As a link technology, the satellite network can connect two computers and provides a point-to-point link. In this architecture, the satellite network is simply a Layer-2 link and is just like other terrestrial links, except having a longer delay. Any Internet traffic can be carried over this link. In today's Internet, satellite networks have provided trunking services for many cross-Atlanta and cross-Pacific Internet links.

As a subnet technology, the satellite network can provide high-speed Internet access for home and office computers. In this architecture, the satellite network is the last-hop access network. Considering its large footprint, this indeed provides a viable "last mile" solution for the Internet. For example, the Hughes DirecPC™/DIRECWAY™ service has provided high-speed Internet access for homes and offices in the United States. Similar products and services are available in Europe and other countries.

In this subnet architecture, the user computers are connected via satellite to a gateway on the Internet. The satellite network has a full set of Layer-3 functions like other subnet technology (such as Ethernet). For example, the gateway may have DHCP function to assign IP address for each satellite network user. It should also support native multicast function. The most significant difference is perhaps the size – a satellite network can potentially have millions of nodes in one subnet.

Finally, as a core network technology, the satellite network can serve as the global tier-1 network for the Internet and carry backbone traffic. Future broadband satellite networks under planning (such as SPACEWAY or Teledesic) all support this role. For example, it is quite convincing that the satellite network can make the best Internet multicast backbone. In this architecture, the satellite network supports and participates in the Internet core routing protocols.

## 2.2 The Role of Satellite in the Satellite Network

Older satellite communication systems are so-called "bent-pipe" because the satellite is a mere signal repeater between two ground stations. There is no data processing on the satellite. From the network viewpoint, the satellite has no role above Layer 1. Bent-pipe satellite systems are simple to deploy but have inefficient use of channel resources.

On-board processing (OBP) is a new type of satellite communication system that allows advanced processing of communication signals on board the satellite. OBP can include modulation, coding, packet switching, routing, and other Layer 2 and Layer 3 functions. Many newer satellite communication systems center on a full data packet switch (such as an ATM switch) in the satellite payload.

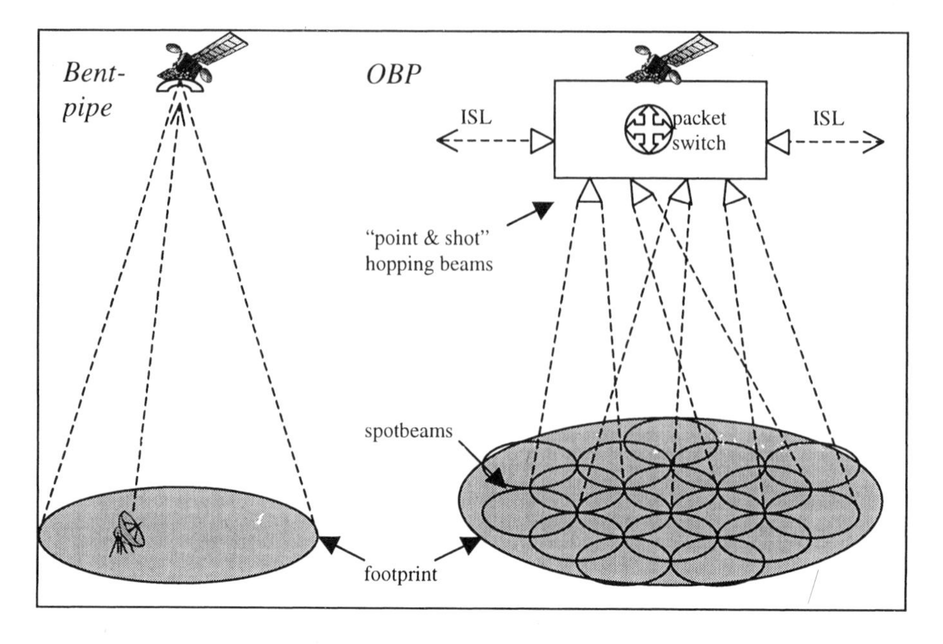


*Figure 1-4*. Satellite roles in a satellite network

Both bent-pipe and OBP satellite networks can be used in providing Internet services. Bent-pipe systems have been used as a link technology and subnet technology (such as in Hughes DirecPC™). A particular type of bent-pipe system is the Direct Broadcast Satellite (DBS), where the communication is broadcast only: from the gateway (uplink) to user computers. To support full duplex Internet access, a separate return channel is used, such as remote dialups through the terrestrial telephone network. The Hughes DirecPC™ network is such an Internet access system based on DBS.

In older satellites, the whole satellite footprint is served in one single beam (the ray of signal transmitted from the satellite to the earth). Newer satellites support multiple hoping spot beams, i.e., a satellite transponder has multiple transmitters and each covers a narrowly focused area. The beams can switch configurations instantly and hop to different areas based on packet destinations, much in a "point-and-shoot" fashion (*Figure 1-4*).

With OBP, the satellite network resembles a switched network such as ATM. Each satellite is a packet switch: switching packets among the receiving beams, transmit beams, and the inter-satellite links. Both Layer-2 and Layer-3 switching have been considered for future satellite constellations. The dilemma of this architecture is the inability to upgrade. Although switching and routing are mature technology, newer capability and faster equipments are being rolled out constantly. Once the satellite is in orbit, it will have to last ten years or longer with the same hardware.

# 3. COMMON APPLICATIONS

Common Internet applications include web browsing, file transfer (FTP), remote login (telnet), video teleconferencing, email, broadcast, etc. Since they all use IP (Internet Protocol) as the transmission mechanism, they can seamlessly run over satellite networks. However, the performance varies among different applications, because their requirements on network bandwidth and responsiveness, their tolerance to communication delay, and the implementation techniques are very different (*Figure 1-5*).

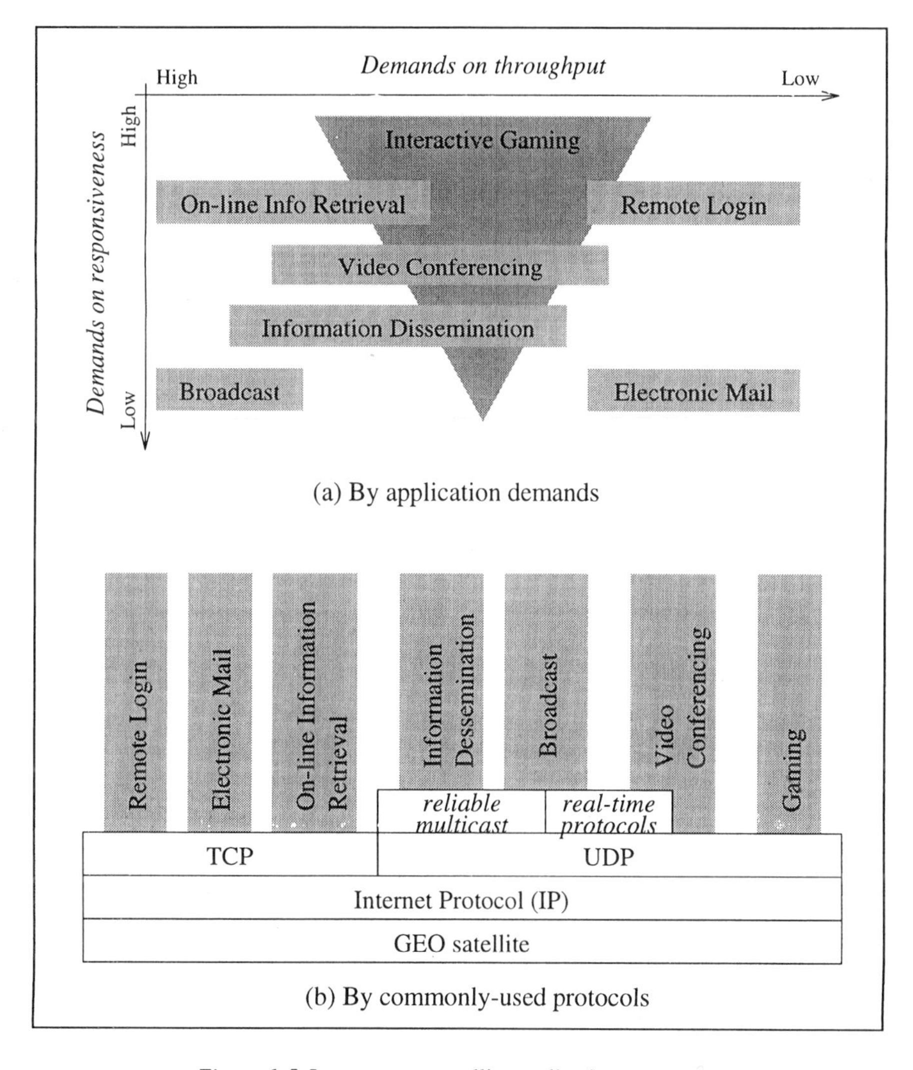


*Figure 1-5.* Internet over satellite application taxonomy

Remote Control and Login

Remote control and login are very delay sensitive. Typically a user expects responsiveness on the order of tens of milliseconds during a remote login session. Remote control may require even faster response, depending on applications. When compared to the often congested and chaotic terrestrial Internet, system response time over GEO satellites is somewhat slower but more stable. If a user can endure half- to one-second delay, remote control and remote login applications can run smoothly over satellite.

Information Dissemination and Broadcast

Satellite networks are better media to deliver bulk data anywhere any time. Some illustrative examples include stock market, financial numbers, and battlefield information. Data broadcast, such as webcasting, network news, and TV programs can be very expensive for point-to-point networks, but is ideally suited to broadcast via satellite networks. New applications such as pushing web cache to POPs or directly to consumers are also broadcast in nature. Therefore, satellite networks are far more suitable for these applications than the traditional terrestrial networks.

Video

Video conferencing and video distribution applications can usually tolerate a certain amount of delay, but little jitter. Typically the protocol requires no bi-directional synchronous (handshake-style) communication, and hence latency is not a prohibitive issue. Compared with terrestrial network, satellite networks can provide better quality in video conferencing due to the available bandwidth and simpler network topology. Jitter can be easily controlled in a satellite network, especially under a single-hop direct-to-consumer architecture.

Electronic Mail/Messaging

Traditionally electronic mail is not interactive. It does not require a great deal of bandwidth and can tolerate reasonable delays (often in terms of minutes). It should work fine over any network.

Information Retrieval (WWW, FTP)

Many such applications require reliable data transmissions and are usually built on top of robust protocols like TCP. Because of the ``acknowledge-and-retransmission" scheme being used, these protocols are often sensitive to the communication latency. However, many of the information retrieval applications, especially the on-line interactive ones, desire the fastest possible response. Their performance will depend on how

TCP is being used in the implementation, i.e., how much information is being retrieved and whether it can be retrieved as a single transfer, or a number of smaller transfers. The issue of TCP over satellite is a big challenge in satellite network research.

Interactive Gaming

Certain applications that require instantaneous reaction time, like highly reactive network gaming, do not work over GEO satellites due to physical delay limitations. Other types of interactive gaming like chess would not suffer from the delay.

# 4. VISIONS FOR THE FUTURE

## 4.1 Commercial Market

Broadband satellite networks have been in planning since the early 1990's; several companies have already launched plans for start of service in the 2003-2004 time frame. According to a recent market analysis report from Northern Sky Research [Baugh 2001], the broadband satellite market is slated for sizable long-term growth. This is driven by the observation that satellites are now recognized as the platform of choice for certain IP applications. According to report, "satellite players are now building business models from the ground-up based on the inherent benefits of satellite technology. Multicasting, global coverage and ubiquity of service will be the core advantaged for satellite companies to leverage as they target lucrative access and content markets."

It is predicted that the global market for Internet-over-satellite will rise from $330 million in 2001 to $12.43 billion in 2006. The global market for multicasting and content delivery services via satellite will also experience sizable growth, from $160 million in 2001 to $3.079 billion in 2006 (see *Table 1-1*).

*Table 1-1.* Global IP via satellite services market (2001-2006) - $billions (source: [Baugh 2001])

| *Satellite Segment* | *2001* | *2002* | *2003* | *2004* | *2005* | *2006* |
|---|---|---|---|---|---|---|
| *Broadband Access* | $0.33 | $1.22 | $3.24 | $6.21 | $9.07 | $12.43 |
| *IP Multicasting and Content Delivery* | $0.16 | $0.29 | $0.52 | $0.94 | $1.70 | $3.08 |

Given this high potential, many companies will venture into this market and compete for their shares. However, this is a risky business considering the sizeable upfront investment and vast technology requirements in order to build and run a satellite network. A few early ventures, including Iridium and ICO, have met with failure and have had to declare bankruptcy. Because the broadband satellite networks will be so much different from the past, a successful venture will likely come from a global consortium of system implementers, operators, and end service providers so as to provide funding, expertise, and far-reaching service capabilities [Cacciamani 2000].

Cacciamani also painted a possible path for the deployment of broadband Internet-over-satellite services. They can start from the market segments that satellites have served well in the past, such as those for the traditional VSAT applications (like retails). They can then expand into multi-media services including video conferencing, distance learning along with major Internet services. Because of the ubiquity feature, satellite networks are good for providing leading edge services that can get to the market quickly. In time, they will also provide services for telecommuters, SOHO, and eventually entertainment services for high-end consumers.

## 4.2 The DARPA NGI Vision

The US administration envisions the emerging information infrastructure to develop into a seamless web of communications network, computers, and consumer electronics and services that will put vast amounts of information at the disposal of its users. Satellites network is one of the major bitways to provide access to the information infrastructure to anyone, anywhere, anytime. To support such emerging information infrastructure, DARPA Next Generation Internet (NGI) provides a vast increase in the geographic scope and heterogeneity of access to the global information infrastructure. Nevertheless, certain applications like digital battlefield, tele-emergency, and multimedia data dissemination often require extended access to remote or rural areas that are outside the reach of conventional communication media such as fiber optics or wireless cellular infrastructure. Satellite networks can supplement this extended coverage of NGI. Such applications require an integration of satellite networks and large mobile multi-hop wireless networks to extend the reach of NGI to support mobile computers deployed for military and civilian applications.

The NGI as envisioned by DARPA NGI program will be a three-tier network structure – the super-high-speed fiber-optic core network, the range-extension sub-networks, and the local access networks. The core network consists of fiber-optic backbones with gigabit to terabit bandwidth. The access networks, such as ad-hoc packet-radio network, serve end-users and concentrate users' traffic of moderate-bandwidth (hundreds of kilobit). The range-extension networks should be the bridge between NGI core and the access network with hundreds of megabit bandwidth. Since global coverage is crucial in NGI, GEO/LEO satellite networks, with their coverage nature, serve as the ideal bridge between the NGI core and the access network.

# 5. CHALLENGES

To summarize, satellite network will be a crucial component of the global Internet. It will complement and bridge the fiber-connected Internet islands and the Wi-Fi or 2G/3G wireless access networks. It will continue to dominate content distribution applications, such as broadcast, streaming, and web cache distribution. Satellite networks will also continue to contribute to overseas trunking, remote access, Intranets, and rapid deployment where there is little or no infrastructure.

We will also face many challenges when we implement the vision outlined above. Any satellite network is a complex engineering. Future success depends on technical breakthroughs in many areas of space technology, data communications, networking, and distributed computing. In the area of space technology, there are many pressing problems ranging from how to coordinate and manage a large number of satellites in a constellation, to how to build smaller and cheaper ground stations. In the area of data communications, researchers are working on OBP, beam forming, media access control and other topics. In the area of networking, issues about integrating satellite networks and terrestrial networks are being worked on. Routing in the LEO satellite constellation, routing between satellite and terrestrial networks, and routing with unidirectional links are all active research topics. TCP-over-satellite is also a networking topic that has gain significant attentions. In the area of distributed computing, there are also many promising research such as how to scale to a potential tens of millions nodes in one satellite network, how to deal with the latency in interactive applications, how to make use of the broadband multicast feature, and how to develop new applications to take advantage of the promising new satellite networks.

# REFERENCE

Cacciamani, G. (2000). *Convergence in the Broadband and Multimedia World*, Hughes Network Systems.

Hu, Y and Li, V. (2001). Satellite-Based Internet: A Tutorial, *IEEE Communications Magazine,* March 2001.

Baugh, C. (2001). *Broadband Satellite Markets: A Comprehensive Analysis of Trends and Opportunities.* Northern Sky Research.

Zhang, Y., DeLucia, D., Ryu, B., and Dao, S. (1997). Satellite Communications in the Global Internet: Issues, Pitfalls, and Potential, *INET'97*.

Chapter 2

# SATELLITE CONSTELLATION NETWORKS

*The path from orbital geometry through network topology to autonomous systems*

Lloyd Wood
*Collaborative researcher, networks group, Centre for Communication Systems Research, University of Surrey; software engineer, Cisco Systems Ltd.*

**Abstract**: Satellite constellations are introduced. The effects of their orbital geometry on network topology and the resulting effects of path delay and handover on network traffic are described. The design of the resulting satellite network as an autonomous system is then discussed.

**Key words**: satellite constellation, network, autonomous system (AS), intersatellite link (ISL), path delay and latency, orbit geometry, Walker, Ballard, star, rosette, *Iridium*, *Teledesic*, *Globalstar*, *ICO*, *Spaceway*, NGSO non-geostationary orbit, LEO low earth orbit, MEO medium earth orbit.

## 1. INTRODUCTION

A single satellite can only cover a part of the world with its communication services; a satellite in geostationary orbit above the Equator cannot see more than 30% of the Earth's surface [Clarke, 1945]. For more complete coverage you need a number of satellites – a satellite constellation. We can describe a satellite constellation as a number of similar satellites, of a similar type and function, designed to be in similar, complementary, orbits for a shared purpose, under shared control. Satellite constellations have been proposed and implemented for use in communications, including networking. Constellations have also been used for geodesy and navigation (the Global Positioning System [Kruesi, 1996] and *Glonass* [Börjesson, et al., 1999]), for remote sensing, and for other scientific applications.

The 1990s were perhaps the public heyday of satellite constellations. In that decade several commercial satellite constellation networks were constructed and came into operation, while a large number of other schemes were proposed commercially to use available frequency bands, then loudly hyped and later quietly scaled back or dropped.

1998 saw the long-awaited launch of commercial services using the 66-active-satellite LEO (low-earth-orbiting) *Iridium* system constructed by Motorola [Leopold and Miller, 1993]. *Iridium* demonstrated the feasibility of Ka-band radio intersatellite links (ISLs) directly interconnecting satellites for wide-scale intersatellite networking. However, *Iridium*'s commercial feasibility was not demonstrated before its operating company had filed for bankruptcy protection. The widespread adoption of mobile telephony and roaming between cellular networks worldwide, largely due to the European GSM standard, had usurped much of *Iridium*'s expected target 'business traveller' market for voice telephony to satellite handsets during the Iridium system's decade-long design and construction period. *Iridium*'s services were later relaunched by a second company, which did not suffer from the original company's need to repay crippling construction debts

The 48-active-satellite LEO *Globalstar* system [Wiedeman and Viterbi, 1993], relying heavily on CDMA-based frequency-sharing technology from Qualcomm, followed *Iridium*, and found the market for a voice telephony service just as difficult. Its operating company filed for bankruptcy protection in early 2002. As the mass market for satellite telephony did not materialise, the focus of *Iridium* and *Globalstar* services was shifted to target niche industrial applications, such as remote mining, construction operations, or maritime and aeronautical use, and low-bit-rate data services (2400bps or 9600bps) were made operational.

Many other proposals looked beyond voice to broadband networking. In 1994 the largest "paper constellation" ever seen was announced; 840 active satellites and 84 in-orbit spares in LEO orbits at 700km altitude for broadband networking to fixed terminals in Ka-band [Tuck et al., 1994]. That proposal was later scaled back by *Teledesic* to a Boeing design of 288 active satellites, which, with its scale and proposed use of intersatellite links, was still more ambitious than the nearest competitor: Alcatel's *Skybridge* proposal for 80 satellites at the same altitude of 1400km [Fraise et al., 2000]. In 2000, *Teledesic*'s parent company took over management of Inmarsat's spinoff *ICO* (for 'intermediate circular orbit'), which had aimed its services at the traditional voice telephony market that *Iridium* and *Globalstar* were designed for, and which had entered bankruptcy protection before even launching [Ghedia et al., 1999].

*ICO*'s mere ten active MEO (medium-earth orbit) repeater satellites without intersatellite links, of which one had been successfully launched and tested by the start of 2002, made for a more realistic, if less exciting, engineering and commercial goal, while *ICO*'s late entry allowed for redesign and increased reuse of popular terrestrial protocol designs, particularly GSM.

## 2. BENEFITS OF GOING TO LEO

The primary advantage that a LEO constellation has over less complex, higher-altitude systems with fewer satellites is that the limited available frequencies that are useful for communicating through the atmosphere can be reused across the Earth's surface in an increased number of separated areas, or spotbeams, within each satellite's coverage footprint. This reuse leads to far higher simultaneous transmission and thus system capacities.

High system capacity was a desirable goal when commercial expectations for the sale of services using those capacities were also high, even though movement of LEO and MEO satellites relative to the Earth's surface means that a number of satellites have to be launched and made operational before continuous coverage of, and commercial service to, an area become possible. (High-altitude balloons and shifts of endlessly-circling aeroplanes carrying transponders have been proposed as a way of increasing frequency reuse while providing lower-cost targeted or incremental deployment.)

As well as being able to provide truly global coverage, LEO and MEO satellite constellations can have significantly decreased end-to-end path delays compared to geostationary satellites, although this is a secondary consideration for many applications. Though free-space loss is decreased by the lower altitude of the satellites, channel, signal and resulting link characteristics are all considerably complicated by rapid satellite movement, the widely varying atmospheric slant path loss as the satellite's elevation changes with respect to the ground terminals that it is communicating with, and by Doppler shift.

Much of the commercial activity in LEO and MEO satellite constellations resulted from a desire to make as much reuse of limited available allocated frequency bands as possible. Frequency allocation is decided globally by the World Radio Congress (WRC), which meets once every two years, and which eventually accepted the legal concept of a non-geostationary satellite service in addition to the already well-established geostationary services.

The United States' Federal Communications Commission (FCC) held a number of targeted frequency allocation auctions in available bands – all the way from L and S up to V-band. This coincided with a flurry of activity in the US aerospace industry and led to a large number of applications for use of those frequencies [Evans, 2000]. The FCC has been the prime mover at the WRC; far fewer commercial constellation proposals have come from outside the US.

Receiving a license for a satellite constellation requires the licensee to commit to launching the described service and using the allocated frequencies by a specified date. If these terms are not met, the license is revoked. Applications have been made for permission to reuse the allocated frequencies terrestrially, demanding changes in the terms of the licenses in order to make it easier to meet and keep them..

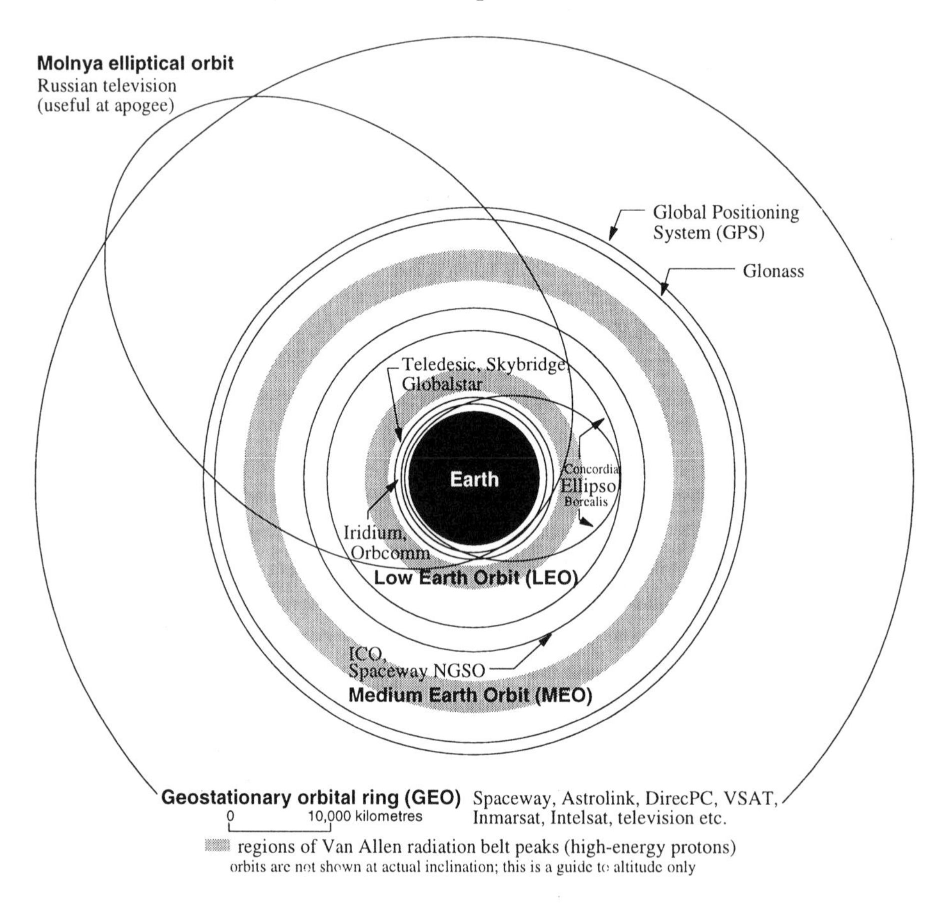


*Figure 2-1.* Orbit altitudes for satellite constellations and proposals

# 3. DESCRIBING THE SYSTEMS

We can categorise satellite constellation networks in a number of simple ways:

- by orbital altitude; LEO, MEO, GEO (geostationary) or HEO (highly elliptical orbits). A brief depiction of existing and proposed satellite constellations is given in *Figure 2-1*.
- by constellation geometry, which is based around satellite positioning and orbit type. This, together with intended service and the limitations of the link budget, determines coverage, which can be regional, targeted or global.
- by frequency bands used for services, from C and L up to Ka and V band, and how this affects the resulting payloads, physical channel and link characteristics.
- by intended service provided by terrestrial user terminals, such as voice telephony, broadband data, navigation or messaging.
- by terminal type. We can group terrestrial user terminals into fixed or mobile terminals. A fixed terminal can be placed and oriented with a permanent view of the sky. A mobile terminal raises roaming issues and increased handover challenges; unlike a carefully-sited fixed terminal, a personal handset can suffer link shadowing and multipath effects that must be considered in the design of the satellite constellation. The power output of a personal handset can also be constrained by radiation limits that are acceptable for nearby humans, and this also affects the overall link budget.
- by the approach taken to implementing networking. Approaches that can be taken to implementing networking range from the simple to the complex. The simpler approaches have separate heterogenous ground networks using passing satellites to complete their radio links. A more complex homogenous autonomous system, built from a space-based network using intersatellite links and smart switching satellites, may peer with terrestrial autonomous systems. This is the fundamental difference in satellite network design approaches between the ground-based *Globalstar* satellite networks and the interconnected *Iridium* satellites, or between the *Skybridge* and *Teledesic* proposals, and is shown using classical network layering in *Figure 2-2* and *Figure 2-3*.

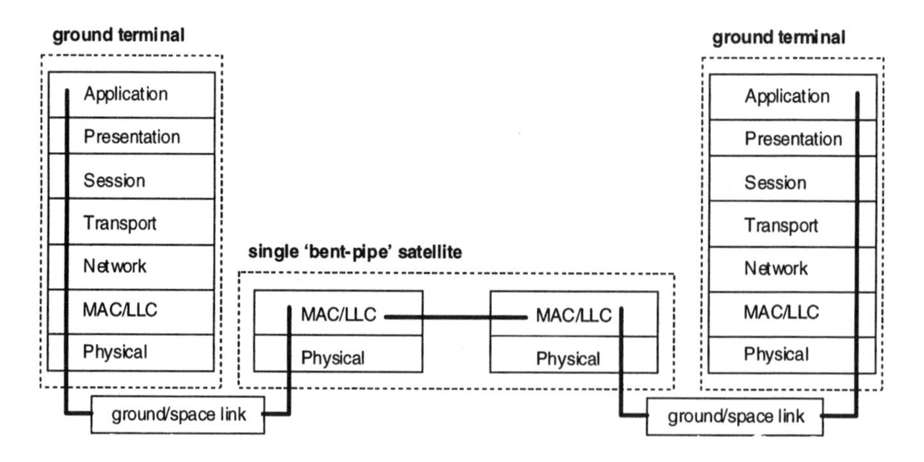


*Figure 2-2.* Repeating satellite approach, e.g. *Globalstar, Skybridge*

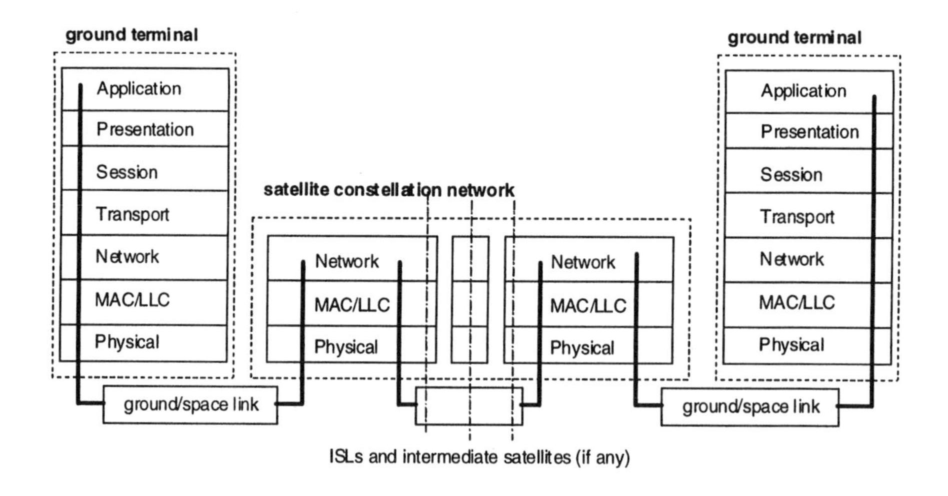


*Figure 2-3.* Full networking and routing approach, e.g. *Iridium, Teledesic*

# 4. GEOMETRY, TOPOLOGY AND DELAY

Orbital mechanics and the resulting satellite geometry have considerable influence over the design of a satellite constellation network. These affect satellite coverage and visibility of satellites available for use by ground terminals, physical propagation considerations such as power constraints and link budgets, and – particularly important from a networking viewpoint – shape the resulting dynamic network topology and the latency of paths across the satellite network. Path latency affects network performance and delay as seen by applications. It is therefore worthwhile to examine the effects of satellite geometry on network topology.

There are a large number of possible useful orbits for satellite constellations. However, preference is given to regular constellations, where all satellites share the same altitude and orbital inclination to the equator, to minimise the effects of precession and simplify control of ground coverage.

Interconnecting a number of geostationary satellites produces a simple ring network around the Equator; an example of this is the geostationary *Spaceway* proposal from Hughes [Fitzpatrick, 1995]. *Spaceway* was later complemented by an additional MEO proposal with intersatellite links, imaginatively named *Spaceway NGSO*, for nongeostationary [Taormina et al., 1997].

At MEO and LEO, the useful types of regular constellation for satellites at the same altitude are generally divided into the categories of 'Walker delta' or 'rosette' [Ballard, 1980] and the 'Walker star' or 'polar' constellations [Walker, 1984]. These are named for the view of orbits seen from above a pole. With intersatellite links, these form variants of toroidal or 'Manhattan' networks [Wood et al., 2001a].

The rosette constellation, where the coverage of satellites in different orbital planes overlaps, provides its best coverage with visibility of multiple satellites from a single ground terminal at the mid-latitudes where most human population lies, but does not cover the poles from LEO. This multiple visibility and availability of multiple physical channels is known as 'diversity'. *Globalstar* uses CDMA recombination of the multiple signal paths between handset and ground station, provided by the overlapping coverage of 'repeater' satellites, to enable diversity to combat shadowing.

At MEO altitudes, *Spaceway NGSO*'s satellites would be sufficiently high to achieve global coverage. The topology of the *Spaceway NGSO* proposal at a moment in time is shown in *Figure 2-4*, where network connectivity between satellites is indicated with straight lines representing intersatellite links. The lighter lines are links between satellites in neighbouring orbital planes. The flowering 'rosette' shape can be easily seen.

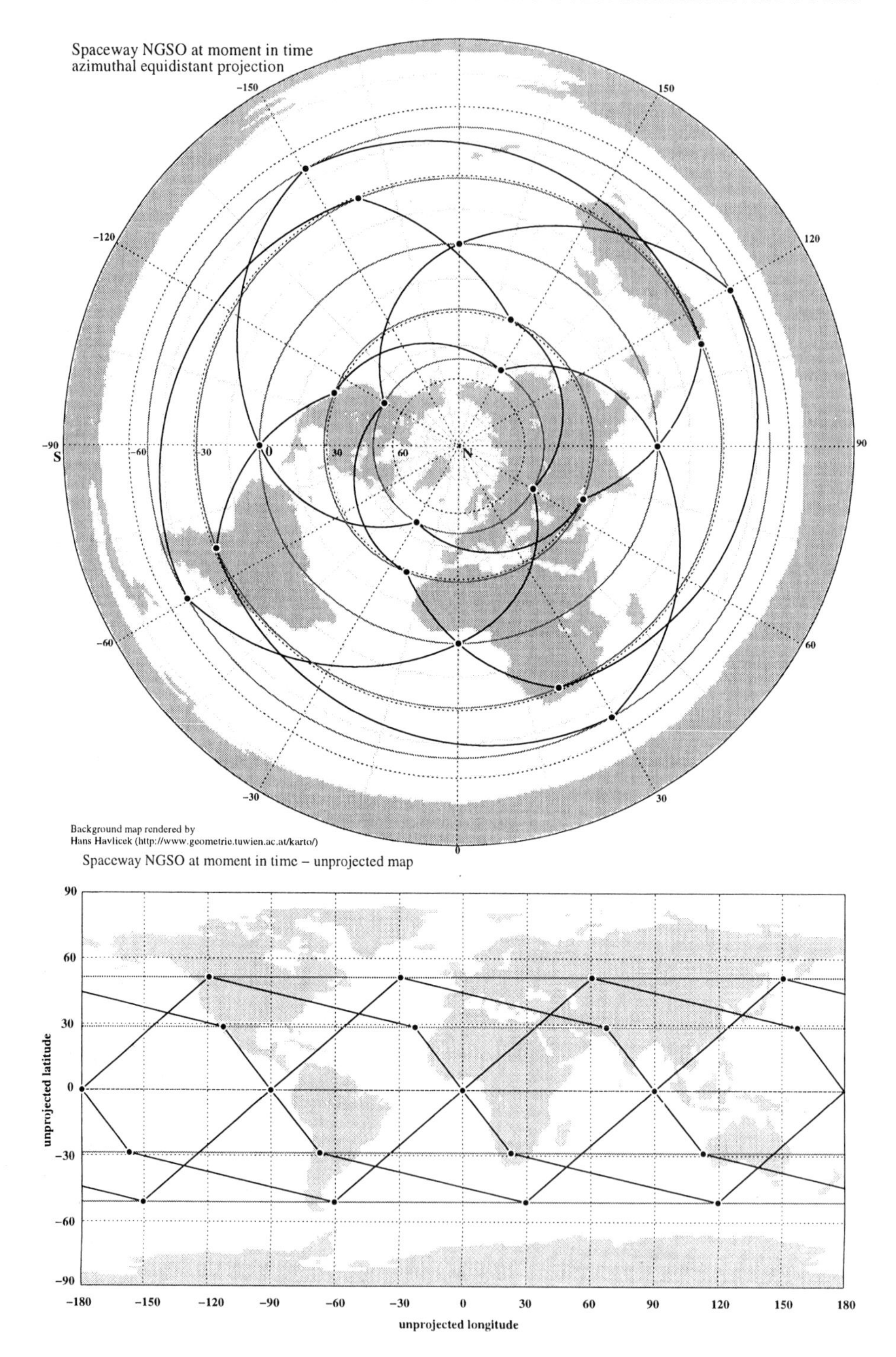


*Figure 2-4.* A rosette constellation: the 20-active-satellite *Spaceway NGSO* proposal

In contrast, the star constellation provides overlapping coverage at the unpopulated poles even at LEO, which is a side-effect of its near-complete global coverage. *Iridium* and the original and Boeing *Teledesic* proposals are based on Walker star geometries. As satellites pass from view they hand over their communication with ground terminals to satellites following them in the same orbital plane, which provides a 'street of coverage' between the similar streets of coverage of neighbouring planes of satellites orbiting in the same direction [Lüders, 1961]. The Earth slowly rotates beneath and across these planes, so that eventually one plane must hand over its terminals to its neighbour to the east. As a result of the Earth's rotation, the 'orbital seam', between the last plane of 'ascending' satellites (travelling north) and the counter-rotating (or 'descending') satellites of the plane almost 180° away, will be encountered by ground terminals. This seam can have a disruptive effect on path delays between terminals and handover between satellites.

Whether the orbital seam between these counter-rotating planes can be spanned by cross-seam intersatellite links that are rapidly handed off between satellites moving at high speeds in opposite directions has been the subject of some debate. With its four intersatellite link terminals per satellite (one fore and one aft to nearby satellites in the same orbital plane, and two to satellites at either side in each co-rotating neighbouring plane) *Iridium* has shown that ISLs work, but its design did not attempt cross-seam communication. The eight intersatellite link terminals per satellite of *Teledesic*'s proposed 'geodesic' mesh would have permitted each satellite at the edge of the seam to maintain one cross-seam link, while the free terminal attempted to establish the next viable link [Henderson, 1999].

Tracking requirements for intersatellite links in LEO star and rosette and MEO constellations and the range of slewing angles required are discussed in [Werner et al., 1995; Werner et al., 1999].

*Figure 2-5* shows the topology of a simulated *Teledesic* proposal at a point in time. The eight-way geodesic mesh of intersatellite links is visible everywhere except at the orbital seam, which has fewer links crossing it. The orbital planes make a 'star' configuration, centred around the pole.

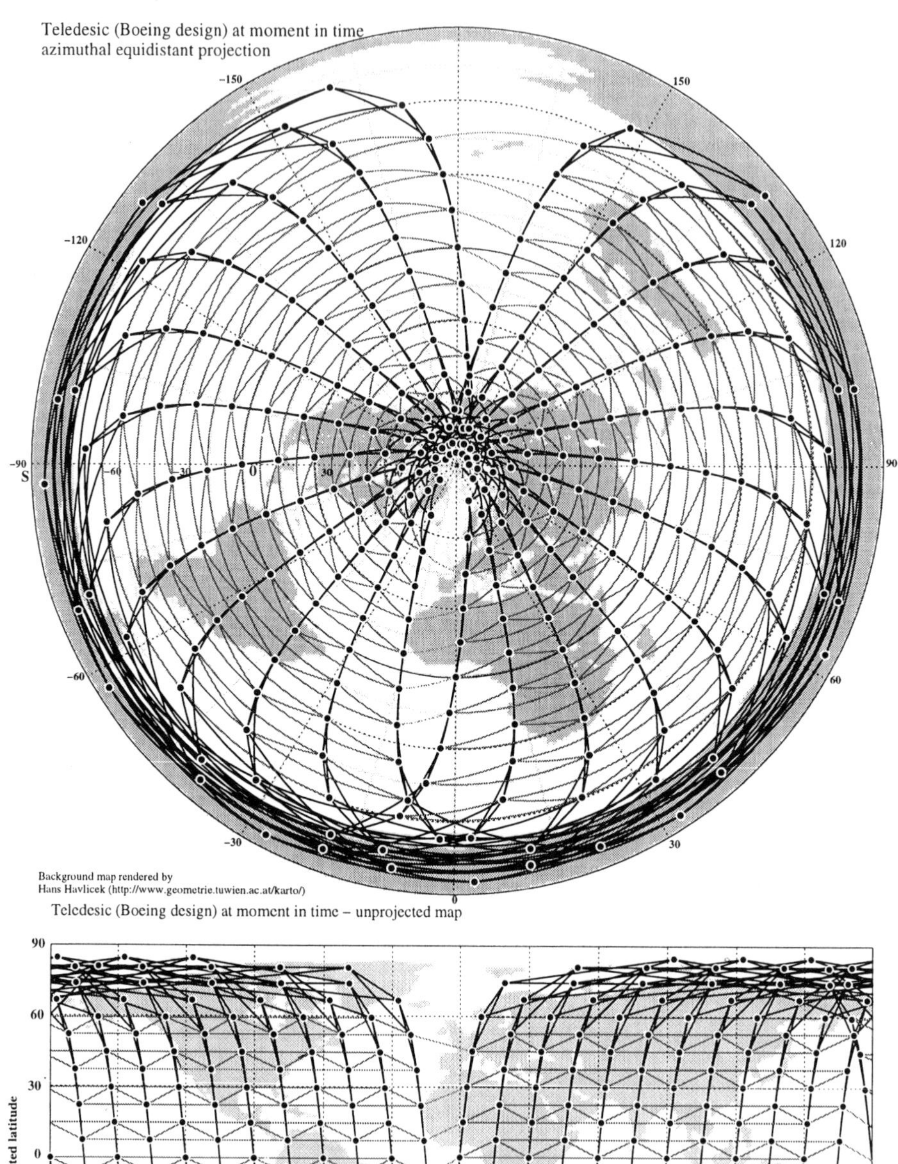


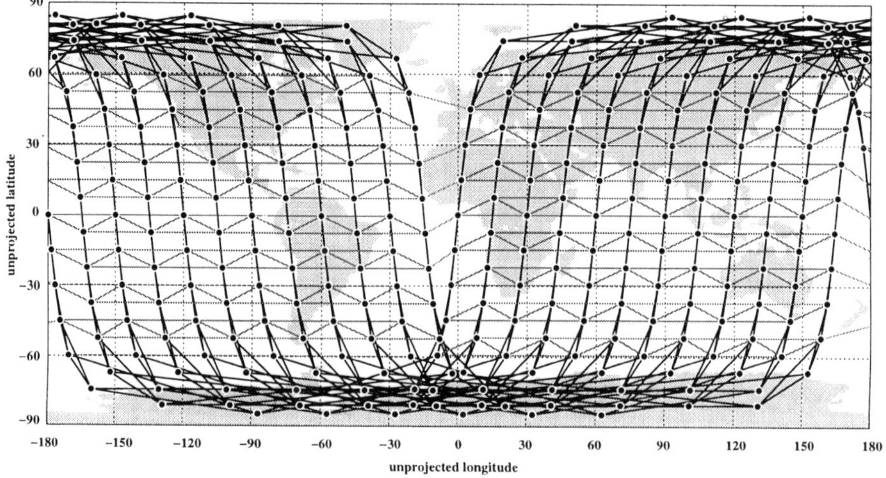

*Figure 2-5.* A star constellation: The 288-active-satellite Boeing *Teledesic* proposal

Constellations have also been proposed using elliptical orbits, where the satellites are only useful while moving slowly at high apogee, and communication link budgets are dimensioned for that distance.

Many useful elliptical orbits are inclined at 63.4° to the Equator, so that orbital motion near apogee matches the angular rotation of the Earth and appears to be stationary with respect to the Earth's surface. Use of *Molnya* (or *Molniya*) and the larger *Tundra* elliptical orbits is now well established for providing broadcast satellite television services targeted to the high-latitude states of the former Soviet Republic; the delay is constant and exceeds the delay to geostationary orbit.

Draim has explored elliptical constellation geometries extensively, and his work was used in the design of the proposed *Ellipso* constellation for voice telephony [Draim et al., 2000].

## 5. DELAY

If a path across a network includes a satellite link, then the delay and error characteristics of that link can have a significant effect on the performance of applications whose traffic uses that path. Tweaking the design of transport protocols to improve their performance when used across the extreme delay and error conditions presented by a link via a geostationary satellite has been a popular field over the years [Postel, 1972; Seo et al., 1988; Partridge and Shepard, 1997; Allman et al., 1999; Allman et al., 2000].

It would be difficult to discuss the error characteristics of the satellite links in a proposed constellation in detail based solely on the constellation topology, since these characteristics are subject to a large number of interrelated engineering design choices at various protocol layers, including antenna design, degree of margin in the link budget, error-control coding choices, and link-layer retransmit strategies for each link. The link and error conditions can also vary over time, as the signal from a ground terminal to a non-geostationary satellite low on the horizon will encounter considerable loss due to the long slant path through the atmosphere. This loss decreases with the shortening of the slant path and increase in signal strength as the satellite rises to its local zenith.

However, the delay incurred by using satellite links is easier to consider and to simulate. For a single geostationary satellite, the coordinates of the ground terminals and the longitude of the satellite are enough to calculate the path propagation delay. For more complex constellations, a first-order approximation can be given by knowing how the constellation geometry and network topology are affected over time by orbital mechanics, and simply calculating the speed-of-light propagation delay between ground terminals at the endpoints of a path. This can be refined by considering the amount of time taken to serialise a frame at a given speed onto the channel at each link. Any jitter due to variations in queuing or switching, or added latency due to contention and link capacity management, will vary considerably according to the specific system design and implementation of the links, and can be considered later.

An example of this propagation delay is given in *Figure 2-6*, for traffic sent from a ground terminal in London, England, to another in Quito, Ecuador, over the course of a day as the Earth completes one whole rotation beneath the planes of a number of different satellite constellations.

The shortest-delay path across each complex mesh network has been selected by routing decisions. As satellite movement is predictable and on a computationally slow timescale, it is possible to predict network topology and handover and to automate updates of routing tables to a considerable degree. Updates can be computed centrally and terrestrially, and then distributed to all the satellites in the constellation by broadcast command. However, handling unexpected link failures gracefully, or engineering traffic flows for quality of service to meet specific application requirements, still requires robust routing algorithms, and has been a popular research area [Mohorčič, et al., 2000; Ekici et al., 2000].

The smallest path delay between two ground terminals is achieved with a LEO constellation, as is shown by the two delay traces for variants on the Boeing-design *Teledesic* proposal. The effect that the orbital seam between counter-rotating planes has on the traffic from one terminal to another is shown by the difference between the two traces. Traffic between the terminals will be rerouted for the several hours that the seam spends passing through the shortest distance on the Earth's surface between the terminals. If there are no cross-seam intersatellite links, the traffic must be rerouted along an orbital plane over the highest latitudes, incurring added delay.

The larger MEO delay, across a simulation of the proposed *Spaceway NGSO* constellation, clearly shows the gradual slow changes in delay due to the cumulative effect of satellite passes from horizon to local zenith and back slowly varying the distances between satellites and terminals, as well as abrupt changes in delay due to handovers between terminals and satellites leading to path rerouting. *Spaceway NGSO* has four orbital planes of five satellites per plane, and these four planes can clearly be seen in the way that the pattern of the delay trace repeats four times over the course of a simulated day, or full rotation of the Earth. As the rosette does not have streets of coverage, we see large alterations in the path delay, where a terminal has handed over communication from an ascending satellite to a descending satellite in a counter-rotating plane, or vice versa. A rosette constellation with large amounts of overlapping coverage can minimise the incidence of such large changes in path delay [Wood et al., 2001c].

With the streets-of-coverage approach taken by *Teledesic* and other 'star' constellations, we only see such alterations in path delay as the orbital seam between ascending and descending satellites intersects the path traffic takes between the ground terminals. Those changes are minimised by the use of cross-seam intersatellite links.

Finally, the two straight lines show the constant propagation delays incurred by using geostationary links. The shorter delay is via a single geostationary satellite at 0°W, while the longer delay is from the London terminal to that satellite, which then uses an intersatellite link to communicate with a second geostationary satellite at 120°W, used by Quito, as part of a three-satellite Clarke constellation [Clarke, 1945]. In practice, allocation of link capacity between multiple terminals using variants on slotted Aloha or other capacity management techniques will lead to longer, more variable delays for network traffic [Maral, 1995].

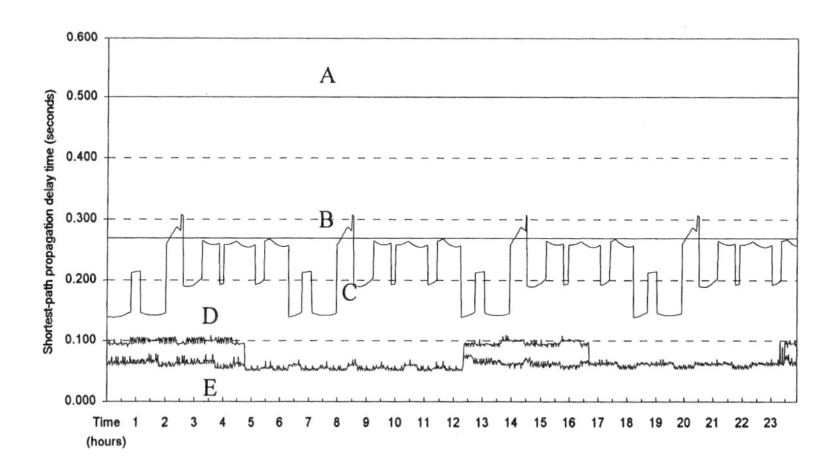


(key to figure: A - hop via two geostationary satellites with an intersatellite link
B - hop via a single geostationary satellite
C - *Spaceway NGSO* proposal
D - Boeing *Teledesic* proposal without cross-seam intersatellite links
E - Boeing *Teledesic* proposal with cross-seam intersatellite links)

*Figure 2-6.* One-way delay between Quito and London via constellations at different altitudes

## 6. HANDOVER

Movement of MEO or LEO satellites, which hand over coverage of ground terminals to other satellites or between multiple neighbouring spotbeams, means that the path taken by traffic between terminals will change over time. When there is a change in path, we can expect changes in path delay.

The path taken will be altered for any packets already in transit whenever terminal handover occurs at the packets' destination. These 'in flight' packets will travel a slightly different path to reach their destination than previous or subsequent packets. This can lead to packet reordering for high-rate traffic, where a number of packets are in flight as handover occurs, resulting in spikes in path delay as handover occurs. The larger distances and propagation delays in the constellation network increase the chances of this affecting in-flight traffic, making the effect greater than in terrestrial wireless networks. This process is shown in *Figure 2-7.*

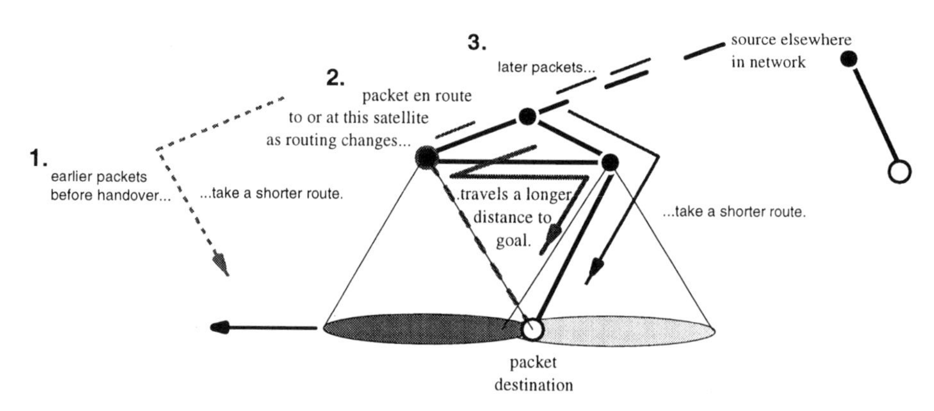


*Figure 2-7.* How handover can affect traffic in flight

Although low-rate traffic is less likely to experience these transient effects during handover, applications sending high-rate and extremely jitter-sensitive traffic can be affected, and the impact of handover on network traffic must be carefully considered in the system design.

If the satellites along the path knew that a handover was about to take place at the destination terminal, it might be possible for them to buffer packets destined to the terminal along the path to prevent those packets from reaching the last hop before downlink until after handover has been completed. However, that would impose a lot of per-flow state on the satellite network, and is not practical for the high-rate traffic that is most likely to experience these transients. Handovers cannot always be easily predicted, particularly for mobile terminals experiencing shadowing.

To illustrate the possibility of transient delay spikes, high-rate traffic was simulated between neighbouring ground terminals. The time between successive packets was less than half of any link propagation delay experienced by the packets. This allowed us to capture and view transient delay spikes due to terminal handover affecting traffic in flight on intersatellite links. These delay spikes would be seen rarely, if at all, by traffic at lower rates.

*Figure 2-8* shows these transients for packets between two terminals communicating using a variant on the *Teledesic* design. The terminals were located on the Equator so that time between plane-to-plane handover was at a maximum, and in similar positions relative to their streets of coverage so that the cumulative effect of satellite passes would be more clearly visible. The steps visible in the curves are due to the resolution of 0.1ms for recorded time within the simulator.

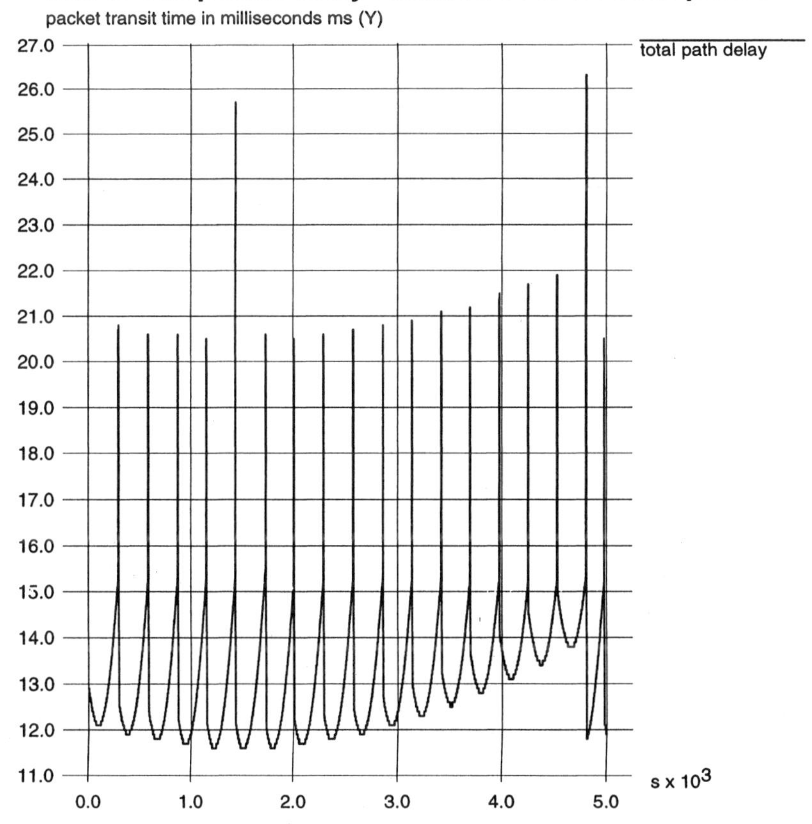


*Figure 2-8.* Path delay for high-rate traffic over a small timescale

Satellite passes, showing a smooth decrease or increase in path delay for packets as satellites approach or leave the terminals' local zeniths, can be clearly seen. Larger path delays for packets in flight during handover between passes are also visible.

These in-flight packets are in the last intersatellite link of their path in the network before the downlink to their destination, or at the satellite the destination terminal is leaving, as that terminal undergoes handover from satellite to satellite. After reaching the satellite previously used by the destination terminal before terminal handover took place, these packets must then be routed onward to the current satellite the terminal is now using. This adds delay before those packets are received. Later packets bypass the former satellite entirely, and the path delay returns to near its previous value.

The overall rotation of the Earth under the planes of satellites, and movement of the terminals across the planes' streets of coverage, showing a gradual decrease and increase in the minimum delays needed to reach local satellite zenith, are also visible in *Figure 2-8* as a great curve that can be drawn tangentially to all satellite pass curves, encompassing the individual passes. These gradual movements under each plane are separated by larger step changes when terminal handover to another plane, and to another street of coverage, takes place. In-flight packets briefly travel two extra intersatellite links to cross the plane, rather than one, before the path changes again as the other terminal hands over as well.

(The large transient spike near 14,000s occurs between the two passes where the uplink and downlink satellites come nearest their terminals' zeniths. The satellites are held onto for the longest period of time before handover occurs at each ground terminal, and the resulting handover is more dramatic as the satellite network has moved the most in its orbits and relative to the surface of the Earth. Since *Teledesic* is a redundantly-connected 'geodesic' mesh, packets can traverse both long and short intersatellite links, of different delays. Here, packets use a long intersatellite link.)

Encountering these transient delay changes by sending a packet just before a handover event occurs could lead to packet reordering, which can adversely impact applications reliant on an ordered flow of packets. The impact of packet reordering on Internet traffic is discussed further in [Wood et al., 2001b].

## 7. NETWORKING DESIGN

How is a satellite constellation designed as a network, and how will it perform? Much existing literature has focused on simulating the performance of a constellation network carrying traditional Poisson telephony traffic [Werner et al., 1997b]. However, doing so involves making assumptions about the nature of the traffic – for example, that the amount of traffic would be large enough for a 'worldwide busy hour' to ensue, where variation in the amount of traffic over the course of a day is insignificant compared to the amount of traffic passing across the constellation network. Such assumptions have not been validated by *Iridium* or *Globalstar*, which have had large amounts of unused capacity.

The rise in popularity of the Internet Protocol (IP) has led to a shift in research focus in the literature from simulating voice telephony traffic to simulating IP data traffic. Simulation of adaptive network traffic, such as that of the Transmission Control Protocol (TCP), part of the Internet Protocol suite, is more complex. TCP's congestion-avoiding slow-start and loss recovery algorithms are affected by the long path delay of geostationary links. However, the movement of LEO and MEO satellites is on a slow timescale, where gradual changes in path delay are often beneath the threshold of notice set by the granularity of TCP timers (typically multiples of 100ms). Simplifying the simulation so that the network path across the constellation is chosen and fixed for a point in time is often sufficient for short simulations of a few TCP flows around that point in time [Wood et al., 2001b].

The adaptive nature of many types of network traffic, when experiencing packet loss due to congestion or to link frame corruption, makes realistic and meaningful large-scale simulation of the network performance of an entire constellation and its traffic difficult. Presuming that the on-board satellite switches are operating as IP routers, with similar functionality and features, also presents a number of problems.

First, IP was originally designed for fixed ground networks, where each network is allocated a fixed address space. Those address spaces, or 'blocks', are then aggregated in routing table entries to allow routers to send traffic to and from each network. Internet routing protocols rely on sensible aggregation of a well-design hierarchy of address blocks to be scalable. However, in a satellite constellation, the topology is dynamic with handover as terminals move between spotbeams and satellites, making it difficult to define and use a stable addressing hierarchy for the terminals. Additions to the IP protocol suite have added needed flexibility to IP addressing – for example, the Dynamic Host Configuration Protocol (DHCP) allows a host to learn of and use an available address that is useful within the local subnet [Droms, 1997]. Mobile IP allows a degree of roaming from a 'home' network for a nomadic IP host moving through other fixed IP networks, using IP-in-IP tunnelling, in order to support remote use of applications which policy has configured for use only on local networks [Perkins, 1998]. However, such additions to the IP protocol suite to allow useful flexible on-demand logical IP addressing for single hosts cannot be used effectively to form the basis of a satellite constellation network, where entire isolated terrestrial networks using a ground terminal for connectivity would need to be fully renumbered with every satellite or spotbeam handover. Mobile IP may handle roaming at the IP addressing level, but without careful integration with lower layers it cannot be considered suitable for rapid handoff on wireless links [Solomon, 1996]. Though a satellite constellation network can be expected to carry significant amounts of both IPv4 and IPv6 traffic across it, it is likely that this traffic will pass through gateways or tunnels on the ground, while the satellites will know how to send encapsulated traffic between ground terminals and nothing more [Wood et al., 2001a].

Second, a dedicated MAC layer is needed to send traffic to individual terminals as they hand over to and from each satellite or satellite spotbeam. The variable size of IP packets does not make a good match with MAC-layer protocols that are most efficient when using fixed frame sizes. Much work has been done on designing wireless ATM for the 'air interface' between ground terminal and satellite, where two ATM cells are wrapped together in a satellite-specific MAC layer frame design complete with error coding checks. A number of proposed constellation designs (including *Skybridge* and *Spaceway*) planned to use an ATM-based MAC frame design. ATM-based routing has been suggested for the constellation network [Werner et al., 1997a], although ATM routing has similar fixed-network assumptions to most IP-based routing.

The satellite constellation network is likely to carry non-IP traffic, including circuit-oriented traffic with strict requirements, and is likely to need traffic engineering to enable smooth handling of path rerouting of ordered flows of traffic due to handovers. The predictive nature of orbital geometry, the need for capacity management of shared spotbeams, and intersatellite and intrasatellite interspotbeam handover for active terminals also mean that engineering optimisations are possible that ad-hoc IP networking cannot consider. The likeliest design outcome is the use of a generic satellite network architecture that is tailored to the physical constellation design and that is able to handle a variety of terrestrial protocols and data types.

The constellation network as an autonomous system will communicate with other autonomous systems. From the outside, the constellation network will appear to be a terrestrial IP network to other terrestrial IP networks, or perhaps an ATM network to other ATM networks. As an IP network it will exchange Border Gateway Protocol (BGP) information with peer IP networks it connects to about the reachability of other IP networks [Rekhter and Li, 1995]. Given the range of terrestrial IP networks that this autonomous system will connect to and know about, and its possible use as a reliable fallback when terrestrial connectivity fails, there is scope for increasing the range of information shared using BGP [Ekici, 2001].

Although the constellation network will carry IP or ATM traffic, that does not make its satellites the equivalent of terrestrial IP routers or ATM switches. Even complex switching satellites within the constellation network are likely to know only how to route traffic between terrestrial terminals and gateways using dedicated interior protocols, and, in the interests of minimising onboard state and complexity, are unlikely to know anything about the world beyond those points. With separation of interior and exterior routing, we can also expect separation of interior and exterior design and network protocol choices that allow a flexible network design to be developed that meets the specific needs of the constellation and supports a variety of traffic types, rather than targeting a single expected traffic type, market and service.

## 8. SIMULATORS

Simulation is an essential tool to gain a solid understanding of the effects of orbital motion on satellite constellations, and how it influences network topology, path delay, and the resulting performance of carried traffic. Some available simulation tools useful for studying satellite constellation networks in a research context are introduced briefly here.

*SaVi* is free satellite visualization software, originally written at the Geometry Center at the University of Minnesota with cooperation from NASA's Jet Propulsion Laboratory [Worfolk and Thurman, 1997]. *SaVi* is useful for showing satellite and coverage movement in two and three dimensions. Although *SaVi*'s source could be extended further to generate handover statistics or compute path delays, its current focus is on showing what would be physically there in a number of well-known proposed constellations. This makes *SaVi* a useful introduction to the effect of orbital mechanics on basic constellation properties, as well as a convenient way of verifying satellite movement in other simulators.

AVM Dynamics has produced the commercial Symmetrical Constellations package, which has a similar focus to *SaVi*, but with generation of relevant delay graphs and statistics. STK produces a commercial package named *The Satellite Toolkit*, whose various modules contain a wide variety of satellite simulation features aimed at industrial use.

*ns*, the network simulator, is free software originally developed within the University of California at Berkeley and now maintained at the Information Sciences Institute (ISI) [Fall and Varadhan, 2002]. *ns* began as a tool to research modifications to TCP algorithms. *ns* still focuses on simulation of changes to TCP/IP, but has now been extended to include types of wireless network. Extensions to simulate satellite movement and the changing link delays of circular orbits were added to *ns* [Henderson and Katz, 2000] and were used for the simulations presented in this chapter. The focus on TCP/IP has traditionally meant far less detail at lower networking layers, although hooks are available that allow specific satellite MAC, LLC and channel layers to be added.

MIL3 has produced the commercial *Opnet* network simulation package, which is also capable of simulating satellite networks, TCP/IP and other existing protocols in detail.

## 9. SUMMARY

Satellite constellation networking can take a space-based or a ground-based approach, with or without the use of complex switching satellites and intersatellite links.

Satellite constellations introduce changes in path delays for network traffic due to satellite movement and handover. These changes can be significant, particularly when traffic is routed across large distances using intersatellite links, and can be modelled in simulation. Regular path changes and handovers form a fundamental part of LEO and MEO satellite constellation networks, and distinguish them from terrestrial networks. Handovers offer particular challenges for high-rate traffic; traffic engineering may be required to keep disruption to traffic to a minimum.

The satellite constellation network will carry IP traffic, but its satellites are unlikely to simply be IP routers; the network design is likely to be tailored to the specific engineering needs of the constellation.

The use of satellite constellations for networking has not yet met the expectations that were raised in the boom years of the mid-nineties. This would seem to make the field more suitable for long-term rather than near-term research, although there has been a decrease in enthusiasm for funding research in this area, just as there has been a decrease in commercial enthusiasm for the satellite industry in general.

## REFERENCES

Allman, M., Glover, D., and Sanchez, L. (1999). Enhancing TCP over satellite channels using standard mechanisms. IETF RFC 2488.

Allman, M., et al. (2000), Ongoing TCP research related to satellites. IETF RFC 2760.

Ballard, A.H. (1980). Rosette constellations of earth satellites. *IEEE Transactions on Aerospace and Electronic Systems*, 16(5):656-673.

Börjesson, J., Johansson, J., and Darin, F., (1999). GLONASS: experiences from the first global campaign. In *Radio Vetenskap och Kommunikation 1999 (RVK '99),* Karlskrona, Sweden.

Clarke, A.C. (1945). Extra-terrestrial relays, *Wireless World*, October, pages 305-308.

Draim, J.E., Cefola, P.J., and Castiel, D. (2000). Elliptical orbit constellations – a new paradigm for higher efficiency in space systems? In *Proceedings of 2000 IEEE Aerospace Conference.*

Droms, R. (1997). Dynamic host configuration protocol. IETF RFC 2131.

Ekici, E., Akyildiz, I.F., and Bender, M.D. (2000). Datagram routing algorithm for LEO satellite networks. In *Proceedings of IEEE INFOCOM 2000*, pages 500-508.

Ekici, E., Akyildiz, I.F., and Bender, M.D. (2001). Network layer integration of terrestrial and satellite networks over BGP-S. In *Proceedings of Globecomm 2001*, pages 2698-2702.

Evans, J.V. (2000). The US filings for multimedia satellites: a review. *International Journal of Satellite Communications*, 18(3):121-160.

Fall, K. and Varadhan, K. (2002). *ns manual / 'ns Notes and Documentation'*, VINT project documentation, available with the network simulator *ns* from http://www.isi.edu/nsnam/

Fraise. P., Coulomb, B., Monteuuis, B., and Soula, J.-L. (2000). SkyBridge LEO satellites: optimized for broadband communications in the 21st century. In *Proceedings of 2000 IEEE Aerospace Conference.*

Fitzpatrick, E.J. (1995). Spaceway system summary. *Space Communications*, 13:7-23.

Ghedia, L., Smith, K., and Titzer, G., (1999). Satellite PCN - the ICO system. *International Journal of Satellite Communications*, Special Issue: LEOs – Little and Big, 17(4):273-289.

Henderson, T.R. (1999). *Networking over Next-Generation Satellite Systems.* PhD dissertation, Computer Science Division, University of California at Berkeley.

Henderson, T.R. and Katz, R.H. (2000). Network simulation for LEO satellite networks. In *Proceedings of the 18th AIAA International Communications Satellite Systems Conference.*

Kruesi, F. (1996). The Global Positioning System: a DOT perspective of where we are and where we are going. In *Proceedings of the Institute of Navigation GPS-96*, Kansas City, Missouri, pages 3-6.

Leopold, R.J. and Miller, A. (1993). The Iridium communications system. *IEEE Potentials*, 12(2):6-9.

Lüders, R.D. (1961). Satellite networks for continuous zonal coverage. *American Rocket Society Journal*, 31:179-184.

Maral, G. (1995). *VSAT Networks*, John Wiley & Son.

Mohorčič, M., Werner, M., Svigelj, A., and Kandus, G. (2000). Alternate link routing for traffic engineering in packet-oriented ISL networks. *International Journal of Satellite Communications*, special issue on the broadband satellite networking mini-conference at IFIP Networking 2000, 19(5):463-480.

Partridge, C. and Shepard, T. (1997). TCP performance over satellite links. *IEEE Network*, 11(5):44-49.

Perkins, C.E. (1998). *Mobile IP*, Prentice Hall.

Postel, J. (1972). Satellite considerations. IETF RFC 346.

Rekhter, Y. and Li, T. (Ed.) (1995). *A Border Gateway Protocol 4 (BGP-4).* IETF RFC 1771.

Solomon, J. (1996). Applicability statement for IP mobility support. IETF RFC 2005.

Seo, K., Crowcroft, J., Spilling, P., Laws, J., and Leddy, J. (1988). Distributed testing and measurement across the Atlantic packet satellite network (SATNET). In *Proceedings of SIGCOMM '88.*

Taormina, F.A. et al. (1997). *Application of Hughes Communications, Inc. for authority to launch and operate Spaceway NGSO, an NGSO expansion to the Spaceway global broadband satellite system*. Filing with the US Federal Communications Commission, Hughes Communications, Inc.

Tuck, E.F., Patterson, D.P., Stuart, J.R., and Lawrence, M.H. (1994). The Calling Network: a global wireless communications system. *International Journal of Satellite Communications*, 12(1):45-61.

Walker, J.G. (1984). Satellite constellations. *Journal of the British Interplanetary Society*, 37:559-571.

Werner, M., Delucchi, C., Vögel, H.-J., Maral, G., and De Ridder, J.-J. (1997a). ATM-based routing in LEO/MEO satellite networks with intersatellite links. *IEEE Journal on Selected Areas in Communications*, 15(1):69-82.

Werner, M., Jahn, A., Lutz, E., and Böttcher, A. (1995). Analysis of system parameters for LEO/ICO-satellite communication networks. *IEEE Journal on Selected Areas in Communications*, 13(2):371-381.

Werner, M., Kroner, O., and Maral, G. (1997b). Analysis of intersatellite links load in a near-polar LEO satellite constellation. In *Proceedings of the International Mobile Satellite Conference 1997*, pages 289-294.

Werner, M., Frings, J., Wauquiez, F., and Maral, G. (1999). Capacity dimensioning of ISL networks in broadband LEO satellite systems. In *Proceedings of the Sixth International Mobile Satellite Conference*, pages 334-341.

Wiedeman, R.A. and Viterbi, A.J. (1993). The Globalstar mobile satellite system for worldwide personal communications. In *Proceedings of the International Mobile Satellite Conference 1993*, pages 291-296.

Wood, L., Clerget, A., Andrikopoulos, I., Pavlou, G., and Dabbous, W. (2001a). IP routing issues in satellite constellation networks. *International Journal of Satellite Communications*, special issue on IP, 19(1):69-92.

Wood, L., Pavlou, G., and Evans, B.G. (2001b). Effects on TCP of routing strategies in satellite constellations. *IEEE Communications Magazine*, special issue on Satellite-Based Internet Technology and Services, 39(3):172-181.

Wood, L., Pavlou, G., and Evans, B.G. (2001c). Managing diversity with handover to provide classes of service in satellite constellation networks. In *Proceedings of the 19th AIAA International Communications Satellite Systems Conference.*

Worfolk, P.A. and Thurman, R.E. *SaVi - software for the visualization and analysis of satellite constellations*. Software developed at The Geometry Center, University of Minnesota. *SaVi* 1.2 is now available from http://savi.sourceforge.net/

Chapter 3

# MEDIUM ACCESS CONTROL PROTOCOLS FOR SATELLITE COMMUNICATIONS

Srikanth V. Krishnamurthy
*Department of Computer Science and Engineering, University of California, Riverside*

Chen Liu
*Department of Electrical Engineering, University of California, Riverside*

Vikram Gupta
*Department of Electrical Engineering, University of California, Riverside*

**Abstract**: This chapter surveys the medium access control (MAC) protocols for satellite networks. Many such protocols have been designed to handle different types of traffic and meet different performance requirements. This chapter gives a comprehensive comparison of these protocols.

## 1. INTRODUCTION

Medium Access Control (MAC) protocols are essential for efficient operations of communication networks. Their main functionality is to arbitrate the access of a shared channel by a plurality of stations in a fair and efficient manner. MAC protocols play a significant role in ensuring the high performance of higher-level protocols such as TCP (Transmission Control Protocol) or IP (Internet Protocol).

In satellite communications, the satellite contains a repeater, called the '*transponder*', between the transmitting and receiving antennas. A transponder operates in parallel on different sub-bands of the total bandwidth used for communications [Maral and Bousquet, 1998]. Communication satellites generally have up to a dozen or so transponders. Each transponder has a beam that covers some portion of the earth. The beam could be a wide beam (10,000 km in diameter) or a spot beam (only 250 km in diameter). Stations on the ground within the beam coverage area can communicate with the satellite on the uplink frequency band. The satellite broadcasts to the stations on the downlink frequency band (different frequencies may be used for uplink and downlink transmissions). Satellites that do not have "on-board" processors, but simply echo whatever they receive from "on-the-ground" processing stations are called *bent pipe* satellites [Tanenbaum, 1996]. In contrast, there are satellites that have on-board processors.

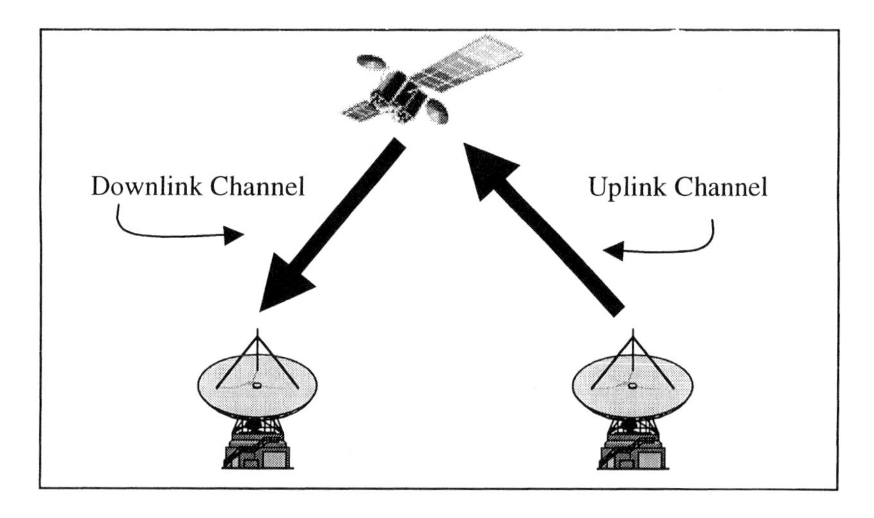


*Figure 3-1.* Uplink and downlink channels in satellite communications

Typically, transmissions on a satellite link can be classified into two types: those on the uplink and those on the downlink (refer to *Figure 3-1*). The uplink transmissions may have a channel that is separate from that of the downlink transmissions. MAC protocols are typically used for the arbitration of uplink access.

A wide range of MAC protocols have been proposed and developed for different operating environments with varying station requirements. In satellite communications, the environment poses some major constraints that eliminate a large number of MAC protocols from consideration. Firstly, the impact of the long propagation delay on the performance of many multiple access protocols (such as protocols proposed for local area and wide area networks [Peyravi, 1999]), limits their applicability to satellite networks. Second, physical changes to the controllers are limited if not impossible (especially if we have an on-board processor) and this necessitates a flexible control mechanism that can be dynamically tuned. The protocol should allow one to easily accommodate changes such as adding or deleting a station and activating or deactivating a station from the network. Finally, a limitation in power necessitates the stringent use of buffer and transponder capacity and the energy consumed by the arbitration process.

Several important factors that affect the performance of a MAC protocol used in a satellite network are listed below:

## Delay Induced by the Environment

The end-to-end delay that a packet suffers has four components. The first is the access delay, which is the time between the packet's arrival instant and the time at which it is considered for transmission. This is simply because a packet is considered for transmission only at the end of the frame during which it arrives. The *frame* is a unit of time predetermined by the underlying MAC protocol. Second is the queuing delay, which is the amount of time that the packet must wait in the queue before being considered for transmission. Third is the packet transmission time, which is equal to the packet size (in bits) divided by the channel bandwidth (in bits per second). Fourth is the propagation delay, which is the time it takes for a bit to traverse the channel or the medium of transmission to reach the intended receiver station. The first three components depend on the packet size whereas the last component is independent of it. While the queuing delays are affected by the access protocol in place and the characteristics of the input traffic, the access delay, transmission delay and the propagation delay are independent of the MAC protocol.

In satellite networks, the propagation delay is the most significant limiting factor on the end-to-end delay that is experienced by a packet. The propagation delay on the station-to-satellite link depends on the satellite orbit. It is small for low orbit satellites (LEOs); however, the relative variations of the propagation delay are large. If we have a geostationary satellite the variations in delay are small. However, the propagation delay itself is significantly longer.

The propagation delay on a station-to-satellite link $t_s$ is given by the following relation $t_s = (R_U + R_D)/c$ where $R_U$ and $R_D$ are the distances from the earth station to the satellite on the uplink and the downlink respectively and $c$ is the velocity of light ($c = 3\times10^8$ m/s). The distances $R_U$ and $R_D$ depend on the location of the communicating stations with respect to the satellite at the time of consideration. For a geostationary satellite, the range over which the propagation time may vary is determined by considering two extreme cases [Maral and Bousquet, 1998]:

(I) The communicating stations are directly beneath the satellite: $R_U = R_D = R_0$, where $R_0$ is the altitude of the satellite ($R_0$ = 35786 km); in this case $t_{SS}$ = 238 ms.

(II) The communicating stations lie on a tangential path from the satellite to the earth: $R_U = R_D = (R_0 + R_E)\cos\theta$, where $R_E$ is the radius of the earth ($R_E$ = 6738 km) and $\theta = \arcsin(R_E/(R_0 + R_E))$. In this case $t_{SS}$ = 278 ms.

The propagation delay that people often use in models is 270 ms. Note that this value holds only for the case of the "on-board" scheduling (OBS) satellite. For the case in which we have "on the ground" scheduling, the propagation delay would be doubled to 540 ms.

It is evident that the maximum and the average packet size impacts the performance of a MAC protocol directly in terms of delay. In addition, the packet size influences the maximum buffer size needed at each station in order to support the offered traffic.

## Centralized vs. Distributed Control

A distributed channel control scheme is one in which channel access decisions are made independently by each station in the system. Thus, each station makes its own decision with regards to the duration and the start times of its transmissions. On the other hand, a centralized control scheme is one in which the channel allocation is done centrally at an arbiter. The arbiter explicitly allocates bandwidth to stations based on certain information (to be discussed) that is provided by the stations. Thus, in a centralized allocation scheme, the stations transmit "requests" to the arbiter, in response to which, the arbiter makes decisions with regards to channel allocation and includes this in "control information" that is transmitted back to the stations. In a distributed scheme, each station disseminates "control information" to other stations (mostly by means of a downlink broadcast via the satellite). Using the control information received from all the other stations, each station can independently decide on when it can access the channel in accordance with imposed policies. The advantages of a centralized access control are the ease of implementation of mechanisms to provide priority access to stations, simpler station logic and a reduced need for group coordination. However, centralized control is inherently less reliable (the arbiter may fail). It may sometimes become the bottle neck with regards to performance in terms of the systems' utilization efficiency (for example, congestion at the arbiter). Furthermore, if the propagation delay is large (which is the case in satellite communications), the distributed control provides significantly lower overhead (in terms of time) than the centralized control. For example, if in a satellite network, an earth station acts as the arbiter, the packet delay would be three times the round trip propagation delay (two round trip delays for obtaining control information from the arbiter and one round trip delay for the transmission of the actual data). This can be reduced if on-board processing is available. In comparison, with distributed control, the minimum delay can be reduced to two round trip delays (only one round trip delay for obtaining control information and another round trip delay for the transmission of data itself) [Tobagi, 1980]. Furthermore, the ability to use the downlink for broadcasting control information in satellite communications makes distributed control a viable option.

## The Traffic Characteristics of the Offered Load

The performance of a MAC protocol depends on the nature of the traffic generated by the individual stations accessing the channel. The traffic characteristics are defined by the pattern in which messages are generated, the length of individual messages and the burstiness of the generated traffic. Traffic burstiness is an important characteristic that influences the design or the selection of a MAC protocol for a satellite network. For bursty stations, MAC protocols using either static assignments or assignments in accordance to reservations (as to be seen in detail later) made over long periods of time are very inefficient. To improve the throughput of a broadcast channel shared by stations with bursty traffic, it is desirable to dynamically allocate transmission capacity on a per message (or packet) basis.

The fundamental objectives, in terms of performance, in the choice of a MAC protocol for satellite communications are:

a) *Achieving high channel throughput*: The throughput of the channel refers to the amount of data that is transported on the channel, on average, in a unit of time. In cases where only a single transmission can be successfully achieved at any given time, the throughput of the transmitting station is the fraction of time in which the channel is utilized for the successful transmission of data of that station.

b) *Achieving low average transmission delay*: The delay of a message refers to the time duration from the moment that the message is generated to the moment at which it is successfully delivered across the channel. A metric often used to characterize network performance is the average delay. This is the long term average of the delays experienced by the messages that are generated in the network. Notice that in the presence of heterogeneous stations (stations that generate different types of traffic), the average delay that is experienced by the messages of a particular station might differ from the average of the delays experienced by all the messages generated in the network.

c) *Ensuring system stability*: A system is said to become unstable if the number of packets queued at any or all of the individual stations grows without bound. In some cases, even though the total message generation rate is smaller than the maximum transmission rate of the channel, the system, still, cannot sustain the message generation rate for long periods of time. Ensuing that the message generation rates are low enough such that the system remains stable is therefore essential.

d) *Facilitating the use of small buffers*: Having to provide larger buffers generally translates into more expensive and complex implementations. A requirement of large buffers usually also means that one might expect longer message delays and vice versa.

There are also other considerations such as ensuring protocol scalability, reconfiguration flexibility in time, low complexity of the underlying control algorithm, providing fairness among all stations and high QoS when applicable, etc., which one cannot overlook when designing a MAC protocol.

MAC protocols that have been designed for deployment in satellite networks may be mainly classified based on the manner in which they perform channel arbitration. We group them into the following categories:

- *Polling based protocols*: Arbiter polls stations in accordance with some predetermined rules.
- *Fixed assignment protocols*: Static allocation of channels to individual stations.
- *Random access protocols*: Protocols that allow individual stations to contend for channel bandwidth.
- *Demand assignment protocols*: Dynamic allocation of channels to individual stations based on their fluctuating needs. These are usually contention free.
- *Hybrid protocols*: These protocols combine reservation and random access methods to improve the efficiency of channel access.

Each protocol has its advantages and limitations. It is seen that there exists no protocol that performs better than all the other protocols over the entire range of performance criteria.

The rest of this chapter will survey MAC protocols in each of these categories.

# 2. POLLING BASED ACCESS PROTOCOLS

One way of arbitrating channel access among competing stations is for a centralized controller to poll them. Having the satellite poll each station in turn, to see if it has a packet to send is prohibitively expensive, given, the 270 ms time required for each poll/response sequence. However, if all the ground stations are also tied to a (typically low-bandwidth) packet-switching network, a minor variation of "the token ring" idea is conceivable [Tanenbaum, 1996]. The idea is to arrange all the stations in a logical ring, so that each station knows its successor. Around this terrestrial ring circulates a token. The satellite never sees the token. A station is allowed to transmit on the uplink only when it has captured the token. If the number of stations on this virtual ring is small and the traffic generated by each station is approximately the same as that generated by every other station, and the token transmission time is short, the scheme is moderately efficient.

# 3. FIXED ASSIGNMENT MULTIPLE ACCESS (FAMA) PROTOCOLS

Fixed Assignment Multiple Access (FAMA) Protocols are contention-free protocols. Contention-free protocols are designed to ensure that a transmission, whenever made, is not interfered with by any other transmission and is therefore successful. This is achieved by allocating channels to stations such that there is no overlap between the channels allocated to different stations. An important advantage of providing conflict-free access is the inherent ability (associated with such schemes) to ensure fairness among stations and the ability to control the packet transmission delay, which is the delay that a packet incurs upon reaching the head of its queue.

In fixed assignment protocols, the allocation of the channel bandwidth to a station is a static assignment and is independent of the station's activities. The bandwidth space is divided into segments and each segment is assigned to a station in a predetermined fashion. In FAMA, channel assignment is not flexible and is not adaptive to traffic changes. This can be wasteful of capacity when the traffic is asymmetric, i.e., if a station generates varying rates of traffic at different times. However, FAMA protocols are suitable for networks where stations transport constant bit rate (CBR) traffic streams for extended periods of time.

FAMA techniques can be classified into Time Division Multiple Access (TDMA), Frequency Division Multiple Access (FDMA) and Code Division Multiple Access (CDMA). One could envision combinations of the above techniques, whereby, a channel is specified by a triplet in frequency, time and code. FDMA was the first technique used in early satellite networks, while TDMA has been widely used in recent years.

## 3.1 Frequency Division Multiple Access (FDMA)

FDMA is the oldest and probably still the most widely used channel allocation scheme [Tanenbaum, 1996; Rom and Sidi 1990; Maral and Bousquet 1998]. In FDMA, no coordination or synchronization is required among stations. A typical 36Mbps transponder bandwidth is normally divided statically into 500 or so 64Kbps channels with each station operating in its own unique frequency channel. Transmissions on one channel do not interfere with transmissions on the other channels.

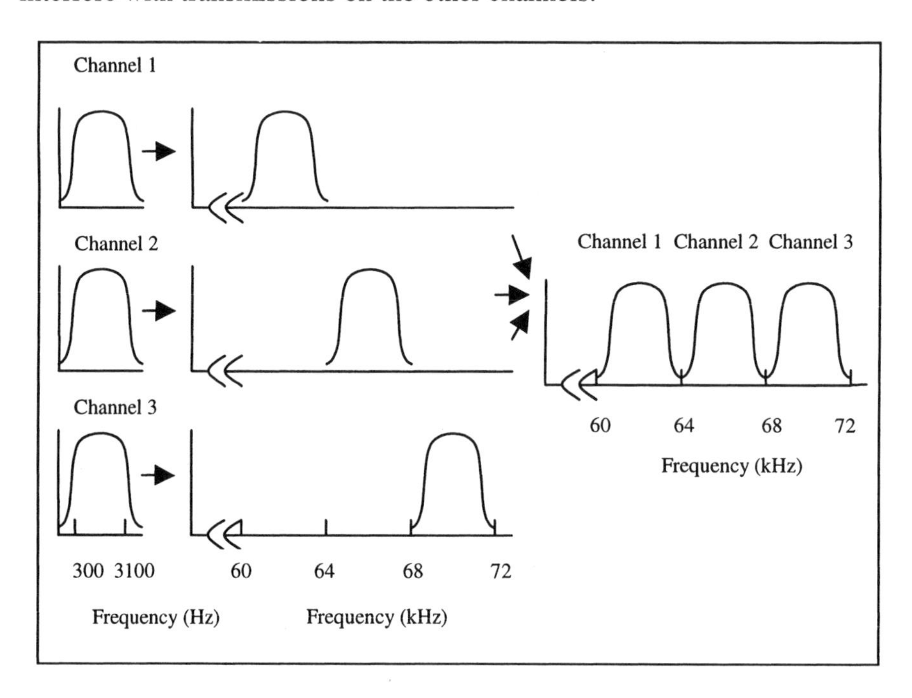


*Figure 3-2.* Channel multiplexing in FDMA

For example, *Figure 3-2* shows how three channels are multiplexed using FDMA. Filters limit the usable bandwidth to about 3000 Hz per channel. When many channels are multiplexed together, 4000 Hz is allocated to each channel to keep them well separated. First the different channels are offset in frequency such that they are disjoint. Thus, no two channels, now, occupy the same portion of the spectrum. Notice that there are gaps (guard bands) between the channels. However, there is some overlap between adjacent channels. This is because of imperfect band-pass filters that are typically used to separate the various frequency bands.

Each station is equipped with a transmitter that operates on its assigned frequency band, and a receiver for each of the other bands (this can be implemented as a single receiver for the entire range with a bank of band-pass filters for the individual bands). FDMA is attractive because of its simplicity; however, it has a few limitations:

a) Guard bands are needed between the channels to keep the stations from causing interference to each other's transmissions (note that even though guard bands are present, as pointed out earlier, interference spills over to adjacent channels due to imperfections in practical filters). The amount of bandwidth wasted due to guard bands can be a substantial fraction of the total bandwidth.
b) Power control is important for FDMA. The transmission power of each station should be controlled to ensure that the interference caused in adjacent channels is not significant.
c) FDMA is an analog technique and does not lend itself well to implementation in software.
d) FDMA can cause a significant wastage of bandwidth especially when the offered load is bursty. When a station is idle, its share of the bandwidth cannot be used by other stations.
e) FDMA comes with a lack of flexibility for reconfiguration. In order to accommodate variations in the capacity allocated per station or in the number of stations, it is necessary to change the frequency assignments. This implies that a modification of the transmitting frequency assignments and the tuning of the corresponding band-pass filters at the earth stations are needed.
f) If the number of earth stations is large, each station gets a very small fraction of the bandwidth.

To summarize, FDMA has the advantage of simplicity, but lacks flexibility and re-configurability. To overcome the drawbacks of static allocation of frequency channels, there are methods that allow the dynamic allocation of channel capacity. We will discuss such methods in a later section.

## 3.2 Time Division Multiple Access (TDMA)

TDMA [Maral and Bousquet, 1998; Peyravi, 1995] provides a better channel throughput than FDMA. In TDMA, the time axis is divided into contiguous slots and these slots are pre-assigned to the stations within the satellite's footprint. Every station is allowed to transmit freely during the slot assigned to it, that is, during the assigned slot the entire bandwidth is available to that station. The slot assignments follow a pre-determined pattern that repeats itself periodically; each such period is called a *cycle* or a *frame*. In the most basic TDMA scheme every station has exactly one slot in every frame (see *Figure 3-3*). There exist TDMA schemes in which several slots are assigned to one station within a frame; these schemes are referred to as Generalized TDMA (G-TDMA) schemes.

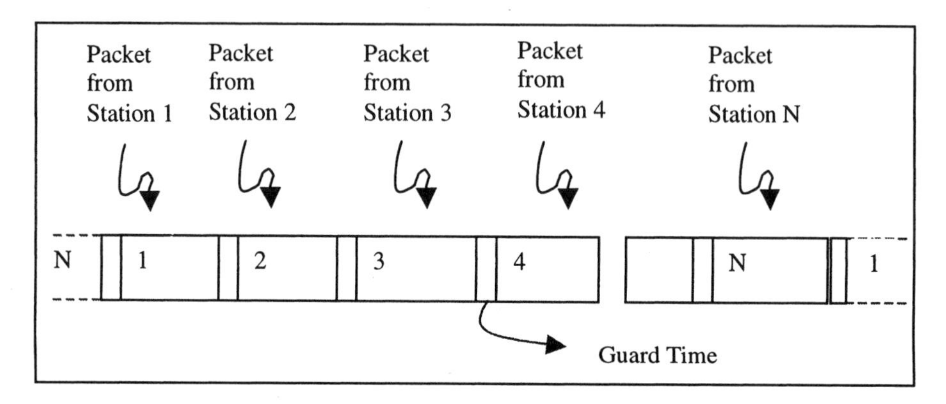


*Figure 3-3.* Slot allocation in TDMA where there are N stations within the satellite's footprint

The advantages of TDMA are as follows:

a) Since only one station is transmitting at a given time, power control is not essential.
b) In FDMA, as the number of stations increases, the number of guard bands increase as well. Thus, a significant fraction of bandwidth is lost due to guard bands when there are a large number of stations. This is not the case with TDMA.
c) Since the frequency band used for transmission is the same for all stations, it is not essential to tune filters in time or upon re-configuration.
d) But, TDMA also has certain disadvantages:
e) The major disadvantage of TDMA is the requirement that each station must have a fixed allocation of channel time whether or not it has data to transmit. Since most applications generate bursty traffic, a fixed allocation of channel time results in wasting precious bandwidth. In a Generalized TDMA scheme, the allocation of bandwidth to a station is proportional to the amount of traffic generated by that station (assumed to be known a priori) and this enables the achievement of a higher throughput.
f) The stations must be synchronized. This is required in order to ensure non-overlapping transmissions. It is not trivial in practice because the

satellite (whose downlink transmissions are often used to synchronize the stations) tends to drift in its orbit, which changes the propagation time to each ground station in time. Although this drift is insignificant as compared with the average propagation delay of 270 ms, it may be significant for synchronization purposes. Guard time may be used (as shown in *Figure 3-3*) to aid synchronization among stations.

g) TDMA also requires each ground station to be capable of transmitting at extremely high speeds. For example, let us consider the Advanced Communication Technology Satellite (ACTS) [Palmer and White, 1990], which was designed for serving a few dozen stations. Even though each of the stations has one 64-kpbs channel overall, since this is a time channel as opposed to being a frequency channel, each station must be capable of transmitting a 64-kbit burst in a 578-μsec time slot. In other words, the transceiver of each station must actually operate at 110 Mbps. In contrast, a 64-kbps FDMA station operates, in reality, at 64 kbps.

When a packet comes to an empty queue, the expected delay in TDMA is equal to half of the length of a frame, since the packet has to wait for the next slot that is allocated to the owner station in order to be transmitted. In FDMA, the expected delay for a packet arriving at an empty queue is equal to the length of a whole frame. This stems from the fact that the actual transmission of a packet in TDMA takes only a single slot while in FDMA it lasts the equivalent of an entire frame. Thus, at low packet arrival rates, TDMA helps achieve lower average delays. However, at high offered loads, when the probability of a packet's arrival to an empty queue is small, the expected delays for the two schemes are both dominated by queuing delays and are approximately equal on average.

Dynamic TDMA is also possible wherein time-slots are allocated dynamically, possibly in response to reservations. These schemes will be discussed in a later sub-section.

## 3.3 Code Division Multiple Access (CDMA)

FDMA does not allow any frequency overlap of the stations' transmissions, while TDMA does not allow any time overlap of the stations' transmissions. A conflict-free protocol that allows simultaneous transmissions, both in frequency and in time, is CDMA [Peyravi, 1999; Tanenbaum, 1996]. With CDMA, stations transmit continuously and together on the same frequency band of the satellite transponder channel. The satellite simply acts as a "*bent pipe*", i.e., it simply echoes the modulated signal from the stations. The principle of CDMA is to modulate the signal from the transmitting station by the addition of a 'signature' which is known to the receiver station and is specific to the transmitting station. This ensures than the receiving station can identify the transmitting station even when many simultaneous transmissions occupy the same frequency band. The signature is in the form of a binary sequence and is most often realized by means of pseudo random codes (Pseudo Noise (PN) codes). The set of codes used must have the following correlation properties:

- Each code must be easily distinguishable from a replica of itself shifted in time.
- Each code must be easily distinguishable from all the other codes used on the network.

Transmission of the useful information combined with the code requires the availability of a much larger radio-frequency bandwidth than that required transmitting the information alone. Hence, CDMA is also referred to as Spread Spectrum Multiple Access (SSMA). The code sequence that serves to spread the spectrum constitutes the 'signature' of the transmitter. The receiver recovers the useful information by appropriately decoding the transmitted signal. CDMA does not require the individual stations to be time synchronized as in TDMA. Furthermore, it does not need channels to be reallocated upon reconfiguration. It is completely decentralized and fully dynamic. The conflict-free property of CDMA is achieved by using orthogonal codes that are detected using matched filters at the corresponding receiving stations. The use of multiple orthogonal signals increases the bandwidth required for transmission [Rom and Sidi, 1990]. Yet, CDMA allows several simultaneous transmissions in the same frequency bands, as long as the different transmissions use different codes. The main disadvantage of CDMA is that the capacity of a CDMA channel in the presence of noise and uncoordinated stations is typically lower than what TDMA can achieve [Maral and Bousquet, 1998].

## 4. RANDOM ACCESS PROTOCOLS

In random access protocols, there is no attempt to coordinate the transmissions of stations so as to avoid simultaneous use of the channel by multiple stations. Instead, each station makes its own decision as to when to access the channel. Random access is simple to implement and is adaptive to varying demand, but it can result in wastage of capacity due to phenomena called *collisions* (to be discussed later). Random access is well suited for networks containing a large number of stations and where, each station is required to transmit short messages infrequently with long idle times between messages.

Early MAC methods that were used for packet satellite systems were based on random access for all packets. In particular, *Aloha* and *Slotted Aloha* were used [Tanenbaum, 1996]. The random access schemes suffer from relatively limited capacity, and they cannot accommodate heavy flows from multiple stations. Furthermore, the long round trip propagation delay aggravates the problem in terms of causing a large increase in latency, since, each packet loss (packet loss is possible and may be quite significant when random access protocols are used due to a phenomenon called collision which we will describe later) adds at least one round trip delay to packet transmission time, because first the loss has to be detected and second the packet will have to be retransmitted. The maximum channel utilization of any random access scheme with a large population (approaching an infinite population) has been shown to have an upper-bounded of 58.7% [Berger and Mehravari, 1981]. We now look at random access protocols in detail. They can be classified into asynchronous and synchronous protocols.

## 4.1 Asynchronous Random Access Protocols

### 4.1.1 Aloha

The basic principle of the ALOHA system [Maral and Bousquet, 1998; Stallings, 1997; Tanenbaum, 1996], or pure ALOHA as it is sometimes called, is to let stations transmit whenever they have data to be sent. After transmission, the transmitting station anticipates an acknowledgement from the receiver station (here we assume that a bent pipe satellite is used). It listens to the channel for a time equal to twice the maximum possible round-trip propagation delay (One round-trip delay is twice the time it takes to send a frame to the satellite) plus a small fixed time increment. If it hears an acknowledgement within that time, it deems that the packet was successfully delivered to the receiver station. If not, it waits for a random period of time and retransmits the packet. If the station fails to receive an acknowledgement after a certain number of repeated transmissions, it gives up.

A receiving station determines the correctness of an incoming packet by examining a packet-check-sequence or a checksum that is included in the packet header. If the packet is valid and if the destination address in the packet matches the address of the receiver station, the station immediately sends an acknowledgement. The packet may be corrupted due to noise on the channel or because of the transmission of another station at about the same time. In the latter case, the two packets may interfere with each other at the receiver so that neither gets through; this is known as a *collision*. If a received packet is deemed invalid, the receiving station simply ignores the packet. The penalty that ALOHA pays for its simplicity is that the maximum throughput of the channel is only about 18% [Tanenbaum, 1996] since the number of collisions and packet retransmissions rise rapidly with increased load.

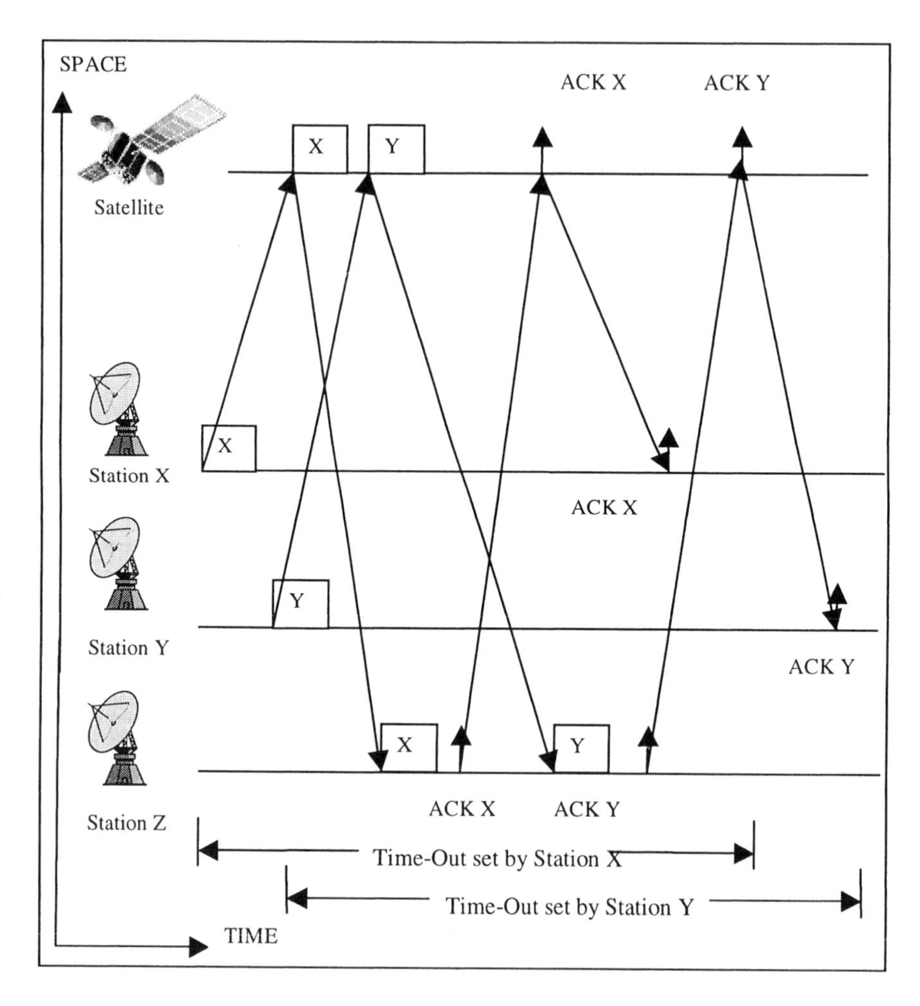


*Figure 3-4.* ALOHA: case when no collisions occur

In *Figure 3-4*, both Station X and Station Y send packets to Station Z. There is no collision and both packets are received correctly. Station X and Station Y also receive an acknowledgement from Station Z within the time-out. These are deemed successful transmissions. In *Figure 3-5*, there is a collision between the packet sent by Station X and the packet sent by Station Y. So neither Station X nor Station Y receives an acknowledgement from

Station Z within their time-outs. Both stations wait for randomly chosen time intervals and transmit the packet again.

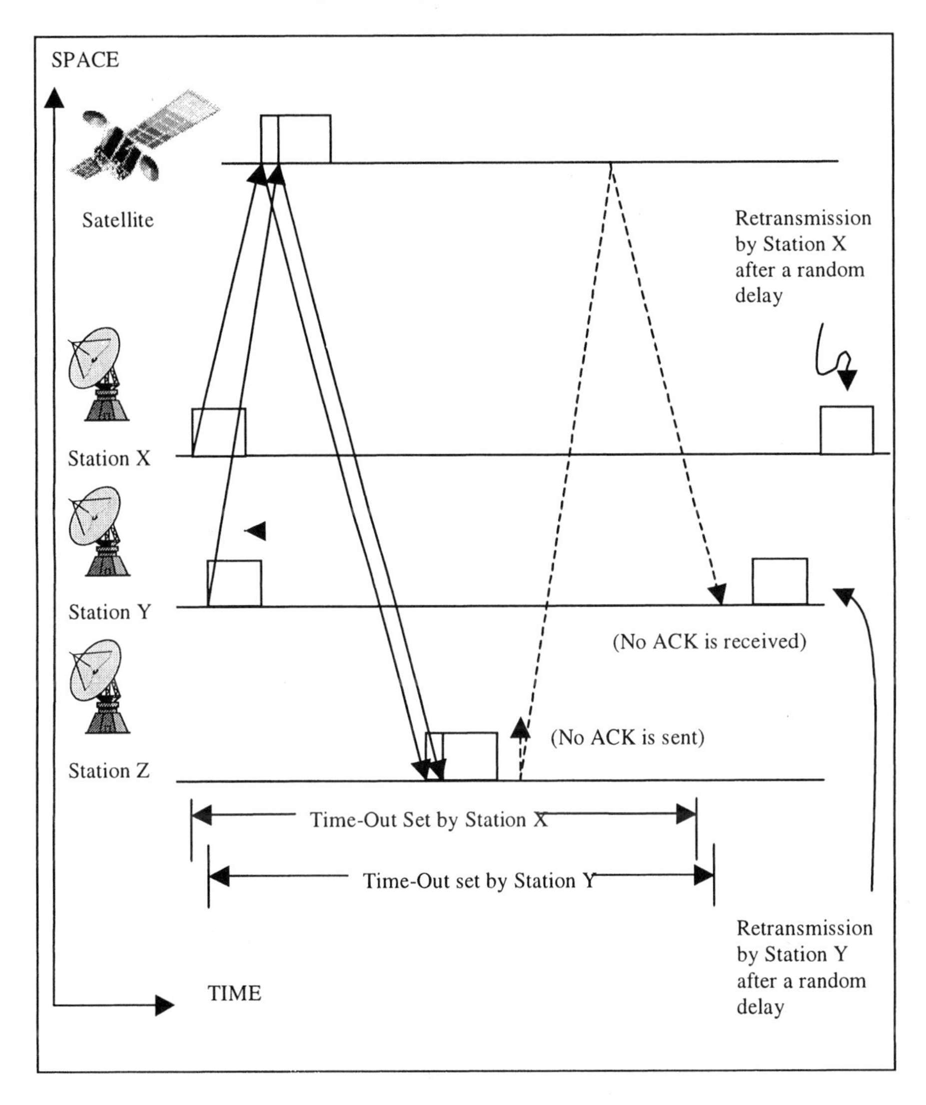


*Figure 3-5.* ALOHA: an example of a collision

### 4.1.2 Selective-Reject Aloha (SREJ-Aloha)

In asynchronous random transmissions, most often, collisions between packets are partial, i.e., two transmissions overlap partially. In Aloha, even if the first bit of a new packet overlaps with just the last bit of a packet that has almost reached the satellite in its entirety, both are totally lost, and both will have to be retransmitted later [Tanenbaum, 1996]. The Selective-Reject Aloha protocol [Maral and Bousquet, 1998] has been designed to avoid a total loss of the packet. Each packet is now divided into smaller sub-packets, each of which has its own header and control bits. When a collision occurs, only the collided sub-packets will be retransmitted (after randomly chosen time-intervals). *Figure 3-6* illustrates the principle of the SREJ-ALOHA protocol. The transmission throughput of the protocol is greater than that of the ALOHA protocol. The practical limit is of the order of 30%. There is some loss in throughput caused by the overhead due to the additional headers that accompany the sub-packets. The Selective- Reject Aloha protocol is well suited for variable packet lengths, since ultimately, the units of transmission are the sub-packets.

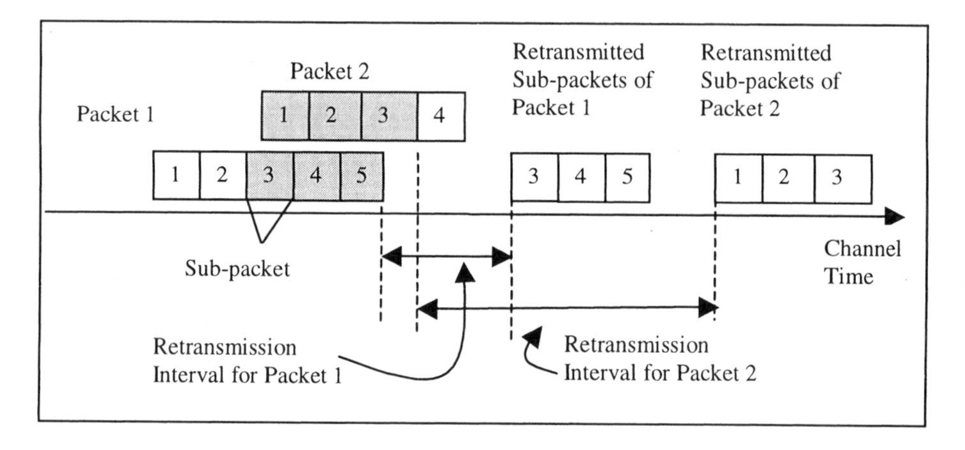


*Figure 3-6.* Splitting packets into sub-packets

## 4.2 Synchronous Random Access Protocols - Slotted Aloha (S-Aloha)

To improve the efficiency of the ALOHA protocol, a modification of the protocol, known as Slotted-ALOHA [Stallings, 1997], was developed. In this scheme, time is divided into contiguous slots. The size of each slot is equal to the packet transmission time. In order for Slotted-ALOHA to function as desired, the earth stations ought to be synchronized in time. One ground station, which may be referred to as the *reference station*, periodically transmits a special signal that is rebroadcast by the satellite and is used by all the other ground stations as a reference in time and to synchronize their clocks. If the time slots are all of length $\Delta T$, each station knows that the time slot 'k', begins at a time k$\Delta T$ after the time origin that is specified by the transmission of the reference station. Since clocks run at slightly different rates, periodic resynchronization is necessary to keep every station in phase. An additional complication is that the propagation time from the satellite is different for each ground station. However, this effect can be corrected for by the use of appropriate guard bands in time. Transmissions are permitted only at slot boundaries. Thus, packets that do overlap will do so completely or not at all. The time scale of the collision is thus reduced to the duration of a packet whereas with the ALOHA protocol this time scale is equal to the duration of two packets as shown in *Figure 3-7*. This divides the probability of the occurrence of a collision by two, which, in turn, increases the maximum throughput of the system to about 37% [Maral and Bousquet, 1998]. It has been shown that constant length packets yield a higher throughput as compared with most other packet length distributions [Bellini and Borgonovo, 1980].

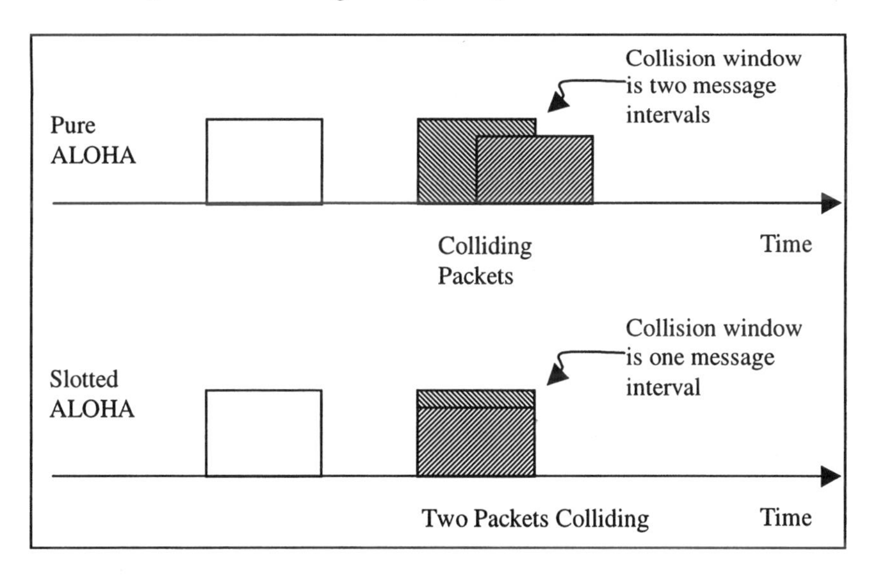


*Figure 3-7.* Examples to show collisions with ALOHA and S-ALOHA protocols

To increase the throughput of the uplink channel to above 37%, we could go from the single uplink channel of *Figure 3-8*(a), to the dual uplink scheme of *Figure 3-8*(b). A station with a frame to transmit chooses one of the two uplink channels at random and sends the frame in the next slot. Each uplink channel then operates independently and on each of the links S-ALOHA is used. If one of the uplink channels contains a single frame, it is just transmitted in a following downlink slot. If both channels contain frames, the satellite can buffer one of the frames and transmit it during an idle slot later on. By doing so, it has been shown in [Tanenbaum, 1996] that the downlink throughput can be improved as compared with that achievable in the case of a single uplink channel. Note that the total bandwidth available (uplink and downlink included) is fixed in the two cases.

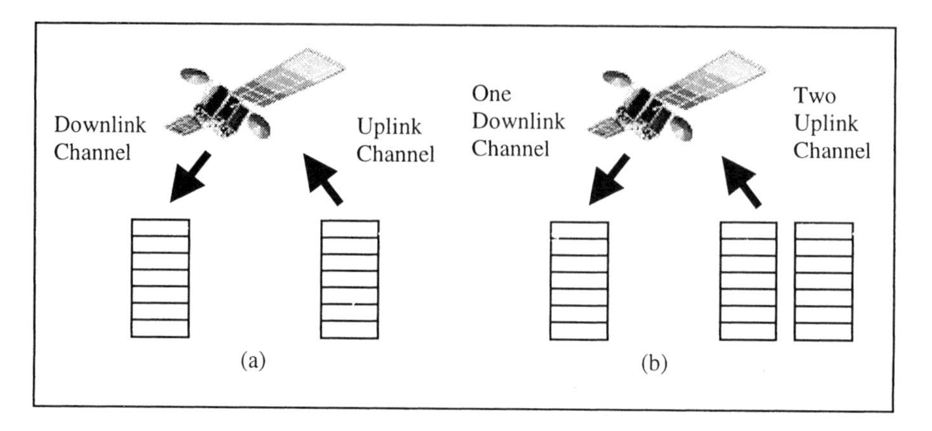


*Figure 3-8.* (a) Standard S-ALOHA (b) adding a second uplink channel

## 4.3 Carrier Sense Multiple Access (CSMA)

Carrier sense multiple access is often used in LANs (Ethernet) to improve the performance beyond what Slotted-Aloha can achieve. However, carrier sensing is impossible in satellite communications since the stations could be potentially far apart and might not be able to overhear the transmission carriers of each other. Hence, CSMA and its variants thereof are not used for satellite communications.

# 5. DEMAND ASSIGNMENT MULTIPLE ACCESS (DAMA) PROTOCOLS

If the traffic pattern is random and unpredictable, a fixed allocation of the channel bandwidth leads to an inefficient use of transponder capacity. Use of random access was seen to provide unsatisfactory low throughput. Thus, it is desirable to deploy MAC protocols that dynamically allocate capacity *on demand* in response to a station's request for capacity. Demand Assignment Multiple Access (DAMA) protocols provide better utilization of the bandwidth when there are a large number of stations requiring low capacity per access, but with large variations in demand patterns. However, while these protocols offer adaptivity and better utilization at high loads, they require additional control overhead that is unnecessary if the static assignment protocols are used. This overhead consumes a portion of the channel bandwidth. Reservation based demand assignment protocols allow an earth station to request the channel arbiter (that might be on-board or on the ground as discussed earlier) for channel bandwidth. In response, the arbiter grants bandwidth to the requesting station. The station can now use the assigned bandwidth to transmit its data without other interfering stations. Dynamic allocation in response to requests based on demand increases the total throughput.

DAMA can be divided into two types of reservation based protocols, those based on *implicit* reservations and those that require *explicit* reservations [Jacobs et al., 1978]. If explicit reservations are needed, a portion of the channel bandwidth is used by stations to make an explicit request for the reservation of bandwidth to transmit one or more packets. The requests for reservations may be transmitted in separate subframe(s) distinct from the frames in which data packets are transmitted, or they may be combined with packet transmissions ("piggybacked" onto data packets) or both. If there are a large number of stations that generate traffic infrequently, explicit reservations may be achieved fairly efficiently with a small number of contention based control slots. The boundary between the control and the data subframes can be made *movable*. Thus, in the absence of information to send, unused portions of the frame time can be converted into control slots that may be used by stations to send in their requests. This reduces the latency incurred in transmitting an explicit request at lighter loads. Control is distributed in that, each station executes an identical assignment algorithm based on global information available on the downlink channel.

The principle behind implicit reservations is what might be referred to as reservation-by-use allocation. Implicit reservations are used in Reservation ALOHA (which will be introduced in a later section). In an example of implicit reservations (refer to *Figure 3-9*), stations use Slotted Aloha to compete for accessing a data slot to start transmission. Stations 1 and 2 transmit simultaneously and experience a collision and thus, neither transmits successfully. Station 3, on the other hand, chooses a unique slot to transmit its packet and hence, is able to make an implicit reservation. Once a station obtains a data slot successfully, it can retain that slot in each frame for its exclusive use. Thus, the particular slot is now, implicitly reserved for this particular station in subsequent frames. Once, a station implicitly receives a share of the bandwidth it could be precluded from participating in the random-access contention process occurring in the remaining slots.

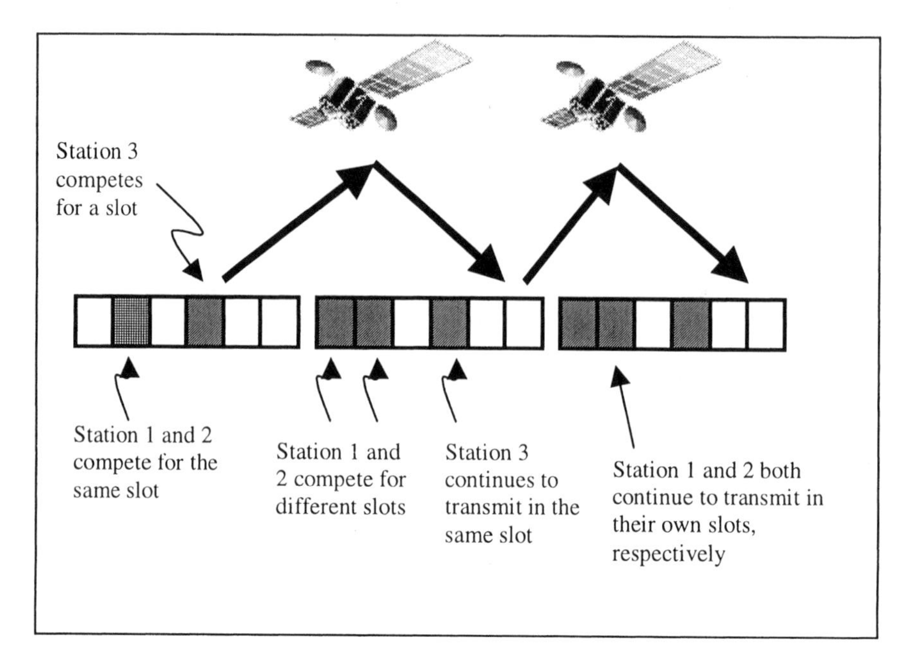


*Figure 3-9.* Implicit reservations

Another possible method for making long-term reservations is by what are called *explicit then implicit* reservations [Le-Ngoc and Yao, 1991]. A station has to first make an explicit request for channel capacity. An on-board scheduler (an on-the-ground scheduler could carry out this functionality as well) will allot channel capacity in response to the explicit request. The allocation of the capacity is implicitly maintained until a channel release message from the pertinent station is received by the scheduler. The explicit then implicit reservation scheme is illustrated in *Figure 3-10.* For simplicity it is assumed that the round trip time is shorter than the frame duration. In this figure we assume the presence of random access slots for making explicit reservations. In the first frame shown, Station 3 is able to make an explicit reservation. It then, holds on to the allocated slot until the fourth frame shown, through implicit reservations. In this frame it releases the reserved slot. Similarly, after an initial collision of requests, Stations 1 and 2 are able to make and then hold on to their reservations. In contrast with this scheme, the R-ALOHA scheme is purely based on implicit reservations. Stations maintain a history of the usage of each slot in a frame (downlink control information is necessary). If a slot is unused, it is open for contention. When a station uses a contention slot successfully, the slot is assigned to that station in each successive frame until it stops using it. The frame duration must be at least one round-trip propagation time or "hop" in this scheme in order to efficiently utilize the channel capacity.

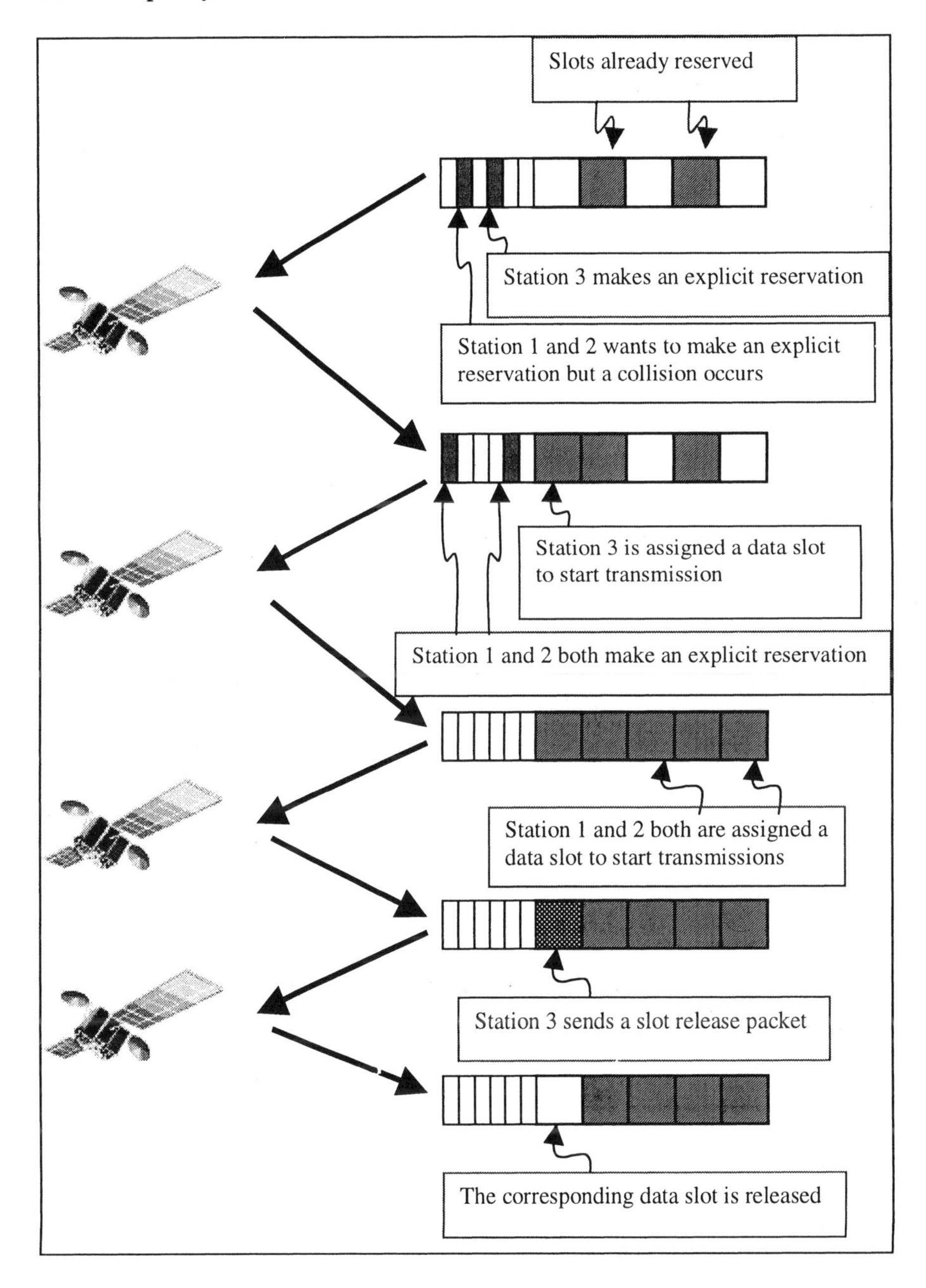


*Figure 3-10.* Explicit then implicit reservation

## 5.1 Demand Assignment Based on FDMA

The SPADE system is an example of a demand assignment system based on FDMA [Maral and Bousquet, 1998; Tanenbaum, 1996]. It is designed to utilize the INTELSAT satellite frequency spectrum of 36 MHz around the 6 GHz band (C Band). The 36 MHz band (refer to *Figure 3-11*) is nominally divided into 800, 45 kHz simplex voice channels. Each channel is capable of carrying 64 kbps digital streams, each of which requires a bandwidth of 32 kHz. The remaining 13 kHz is used for providing a guard band. The voice channels are used in pairs to provide full duplex service. In addition there is a 160 kHZ common signalling channel (CSC) located at the bottom of the spectrum (overlapping with Channel 1) as shown in *Figure 3-11*. This provides a 128 kbps link used for the purposes of facilitating signalling between earth stations. Access to the CSC is in accordance with a TDMA scheme. The common signalling channel is divided in time into 20 units of 50 msec each. A unit contains 50 slots of 1 msec (128 bits) each, and is allocated to specific ground stations. Of the 800 voice channels, only 794 simplex voice channels are used and the other 6 channels provide a separation between the 397 uplink channels and the 397 downlink channels, and between the simplex voice channels and the CSC. When a ground station has data to send, it picks a channel that is currently unused, at random, and indicates the number of that channel in its next 128-bit slot. If the selected channel is still unused when the request is responded to on the downlink, the channel is considered allocated and all the other stations are precluded from attempting to acquire it. If two or more stations try to obtain the same channel at about the same time, a collision occurs and they will have to try again at a later time. When a station is finished using its channel, it sends a de-allocation message in its slot on the common channel. There is an advantage in using the dynamic frequency allocation as in the SPADE system over the static FDMA scheme when there are a large number of earth stations with bursty traffic. The penalty paid in using the SPADE system is the cost of additional processing required at each earth station to facilitate the demand assignment.

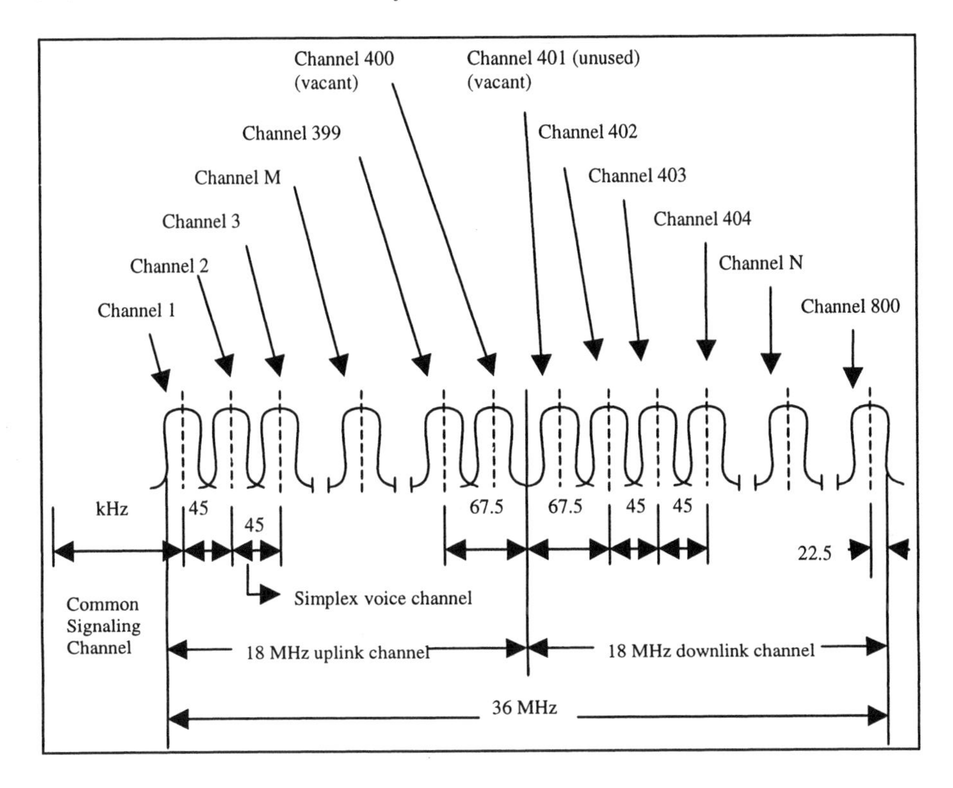


*Figure 3-11.* Frequency multiplexing in INTELSAT SPADE

## 5.2 Making Reservations by Contention Based Access

The objective in using reservation protocols is to avoid collisions entirely. Since stations are distributed, a reservation sub-channel (or control channel) is necessary for stations to communicate with the satellite to make reservations, such that, only one station can access the channel at a time during actual data transmissions. For making reservations most protocols adopt either a fixed assignment based TDMA protocol or some variation of the S-Aloha protocol. A TDMA protocol performs poorly if there are a large number of stations. On the other hand, the S-Aloha protocol is independent of the number of stations, but it needs to be adaptively controlled for stable operations.

### 5.2.1 Reservation Aloha (R-Aloha)

R-Aloha [Lam, 1980] is a distributed protocol combining random access and demand assignment. It facilitates a means of making implicit reservations. Channel time is divided into contiguous frames of equal size and the length of each frame spans the duration of one round-trip propagation delay. Each frame contains multiple slots. The slot size is equal to the transmission time of a single packet. It is assumed that the packet sizes are of constant length. If the frame size in time is not at least equal in duration to one propagation delay, a slot could go unused multiple times successively before stations sense its availability. Reservations are implicit in the sense that a successful transmission in a slot serves as a reservation for the corresponding slot in the next frame. The satellite has to indicate on the downlink as to which slots are open and which are not. A slot is deemed unused in a frame either if it was unused in the previous frame, or if more than one user attempted to transmit in that slot in the previous frame as a consequence of which, there was a collision. Initial access is by using S-ALOHA (random access). A transmitting station contends for a randomly chosen slot for uplink access. Once a slot is successfully accessed by a station, the "same slot" in the succeeding frames is reserved for the same station as long as it has data to send.

R-ALOHA works well for both bursty and constant bit rate (CBR) traffic applications. If the traffic is CBR or if large volumes of data are to be transferred, the implicit reservations help in reserving bandwidth continually. For short messages that fit into a slot, random access (S-ALOHA) is used. R-ALOHA is unsuitable if the traffic simply consists of short messages that fit into a single slot; since it takes one round trip time to determine the slots that are unused (and hence, open for contention), an access latency of 270 ms is incurred. Thus, for such offered traffic one might expect the performance of R-ALOHA to be worse than that of S-ALOHA as a result of the fact that slots used successfully have to be unused for one subsequent frame before they are open for contention again.

The advantage of R-ALOHA is that it can efficiently handle bursty data traffic. However, the disadvantages are that it is 1) inefficient for single packet messages; 2) requires tracking and downlink broadcast of the status of slots with regards to whether they were used or unused in previous frames; 3) frame sizes could be excessive since a desirable feature is that the frame size be larger than a "hop" delay.

### 5.2.2 Priority-Oriented Demand Assignment (PODA)

Priority-Oriented Demand Assignment or PODA [Jacobs et al., 1978] is a demand assignment protocol designed to efficiently support both datagram and CBR traffic. It is also designed to support multiple classes of traffic with varying delay requirements and priorities. It is also expected to support variable message lengths. Its design is based on an integration of multiple demand assignment and control techniques to support both circuit and packet-switched connections.

a) Explicit reservations are used for datagram messages. This results in a propagation delay component of at least two satellite hops, once for the reservation and once for the message itself.
b) A single explicit reservation is used to set up each CBR stream, with subsequent packets of the stream receiving automatically scheduled reservations at predetermined fixed intervals. Thus, subsequent packets experience just one-satellite-hop delay instead of two.
c) Very high reliability, when desired, can be provided by the use of scheduled acknowledgements and packet retransmissions.

Channel time is divided into two subframes as shown in *Figure 3-12*, an information subframe and a control subframe. The information subframe is used for the transmission of scheduled datagrams and CBR streams. These packets can also contain implicit control information such as piggybacked reservations and acknowledgements in their headers. Under normal operations, the control subframe is used to send explicit reservations if implicit reservations cannot be sent in the information subframe in a timely manner.

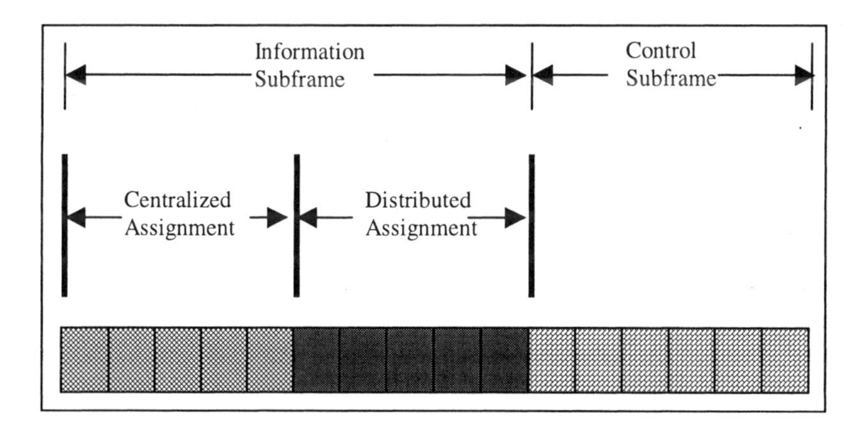


*Figure 3-12.* PODA frame structure

The control subframe must be used for initial reservation access by a station which has no impending scheduled transmissions. It may also be used to make reservations for a delay-sensitive traffic stream if the station cannot wait for the next opportunity to use the header of a scheduled message to make the reservation for bandwidth for that stream. The access method used for this subframe depends on the characteristics of the system in which PODA is being used. If the total number of stations is small, then a fixed assignment of one slot per station is typically employed within the control subframe and the protocol is referred to as FPODA (for Fixed PODA). For large populations of low duty-cycle stations, or in situations in which, traffic requirements are not well known, random access (S-ALOHA) is used for accessing the slots in the control subframe and the protocol is referred to as CPODA (for Contention-based PODA). Combinations of fixed assignment and random access to arbitrate access in the control subframe are also possible. In either case, in this subframe time is divided into fixed contiguous slots, each of which is equal in length to the size of a control packet of fixed length.

An explicit reservation is made for each datagram messages generated at an owner station and the datagram is inserted into a scheduling queue (if a slot in the future is allocated as indicated by the satellite on the downlink) that is maintained by the owner station. The station performs a distributed scheduling of its flows in accordance to their relative importance. The contents of the queue are ordered in accordance with the *reservation urgency*, which is a function of the potential latency or deadline (as determined by the "application-specified" delay class) and the relative priorities of the datagram messages. Messages that are of equal importance (as specified by the latency and/or priority criteria) are further ordered to provide fairness to the applications involved. Channel time allocated in the information subframe is then used to transmit messages as per the ordering in the scheduling queue. Notice that the response to a request for a given message can only be obtained after a delay of one round-trip time if there is an "on-board" arbiter (or two round-trip times if we have an "on the ground" arbiter). Thus, it is possible for a higher priority message, arriving in the meantime, to be preemptively transmitted in slots that were reserved due to prior lower priority messages.

A reservation for a CBR stream is made only once, at inception. CBR streams are assumed to be generated in periodic repetitive bursts. Each stream has specific parameters such as the repetition interval, the maximum tolerable delay relative to this interval and a priority associated with the stream. At each instance of the repetitive interval, a newly generated message from stream is inserted into the scheduling queue. This message is then treated as if it is a "datagram" and in accordance with the scheduling rules. It is possible for higher priority datagram traffic to preempt lower priority CBR streams for indefinitely long durations. This effect could even result in the termination of a low priority CBR stream if it is unable to meet the delay requirements.

To maximize channel efficiency while satisfying specified delay constraints, a reservation may actually be made for an aggregation of several distinct datagrams. We call such a reservation a *group reservation*. The maximum number of datagrams that may be allowed per group reservation represents a trade-off between satisfying the "system urgency constraints" for competing stations (allowing the aggregation of a large number of datagrams might allow the domination of the channel by a single application, thereby violating the urgency requirements of other applications) and the reduction in the relative burst overhead that is achieved as a consequence of aggregating a larger number of datagrams. Thus, small messages are packed into a single large packet when possible, as shown in *Figure 3-13*.

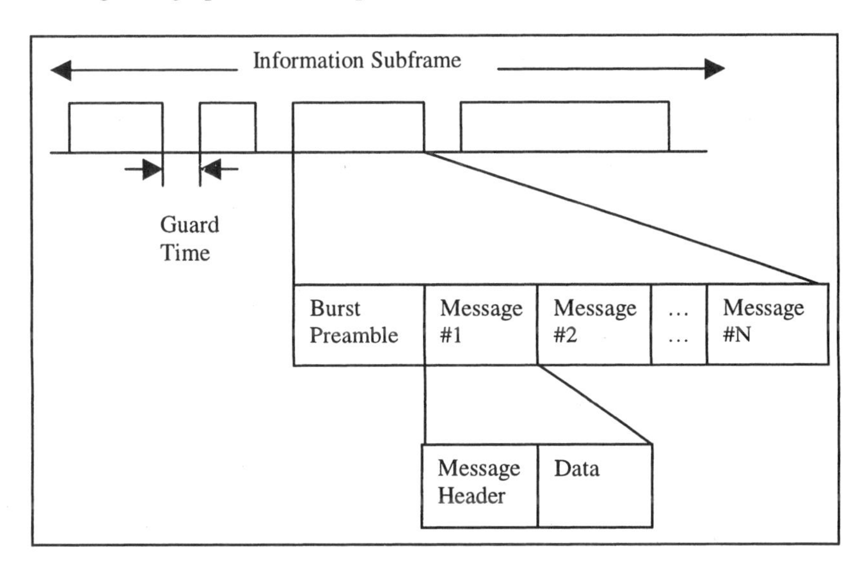


*Figure 3-13.* Aggregating messages to reduce preamble overhead

If reliable communications are necessary, a receiver has to send an acknowledgement to the sender for each received packet. It would then have to make reservations in order to actually transmit an acknowledgement. Reliability may be improved by using error correction codes.

The satellite channel efficiency achievable with PODA is clearly limited by the average control subframe size and the overhead used for making reservations in the information subframe. The latter is strongly dependent on the average size of aggregated burst that is transmitted in sequence. The relative overhead is small even for short packets since a request packet is small. The choice of the size of the control subframe determines the trade-off between channel efficiency and the performance in terms of improved delay and fairness. An important aspect of PODA, relative to other demand assignment approaches, is the processing requirements at each station. The requirement in terms of continually scheduling packet transmissions could potentially be processor intensive, especially if we have a high data rate channel (The higher the data rate of the channel, the faster the scheduling ought to be).

### 5.2.3 Split-Channel Reservation Multiple Access (SRMA)

Split-Channel Reservation Multiple Access or SRMA [Tobagi and Kleinrock, 1976] is a centralized protocol based on explicit reservations. The available bandwidth on the uplink is divided into two channels. One of these channels is used to transmit control information; the other is used for transmitting the messages themselves.

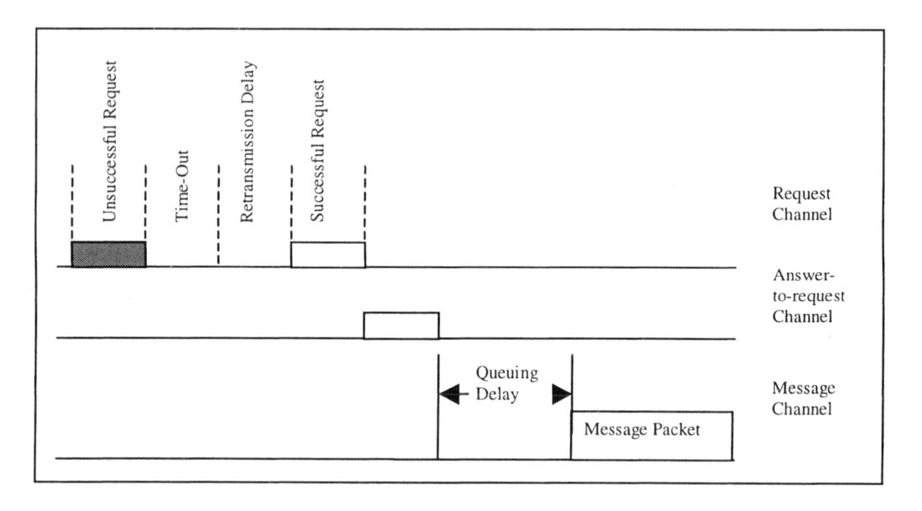


*Figure 3-14.* SRMA: the RAM version

There are multiple versions of this protocol. In the version called the 'request, answer-to-request message' (RAM) version, the bandwidth used for the transmission of control information is further divided into two channels, namely the request channel and the answer-to-request channel (refer to *Figure 3-14*). The request channel is operated in a random-access mode (ALOHA or S-ALOHA). To initiate the sending of a message, the owner station sends a request packet containing the address of the destination station, and in the case of variable length or multi-packet messages, the length of the message, on the request channel. Upon the correct reception of the request packet, the arbiter determines (based upon the prior requests that have been received) the time at which the requesting station can begin its transmission and sends a response message to that station on the answer-to-request channel, indicating this time. At this allocated time, the originating station would then transmit the actual data.

In another version of SRMA, called the RM (request-message) version, the total available bandwidth is divided into two channels only: the request channel and the message channel. As before, the request channel is operated in a random-access mode. A station with a message ready for transmission sends a request packet (containing its address) on the request channel. When this request is correctly received by the arbiter, the request packet is inserted into a request queue. Requests may be served on a FCFS (first come first serve) basis or in accordance with any other scheduling algorithm. A response packet containing the address of the queued station which had issued the request at the head of the queue is transmitted by the arbiter on the downlink. After hearing its own address being announced by the arbiter, the station starts transmitting its message on the message channel. If a station does not hear its own address being announced by the arbiter within a certain appropriate time after the request was sent, the original transmission of the request packet is assumed to be unsuccessful. The request packet is then retransmitted. The process is repeated until a response is heard or a pre-determined number of retransmission attempts are made, upon which the packet is discarded.

### 5.2.4 The Time-of-Arrival Collision Resolution Algorithm (CRA):

In the Time-of-Arrival Collision Resolution Algorithm or the CRA protocol as it is referred to [Maral and Bousquet, 1998], the channel time is again divided into contiguous frames. The length of the frame in time is equal to one round-trip propagation delay. Each frame is further divided into two subframes, the reservation subframe and the information subframe. The information subframe is divided into slots, the length of each of which is equal to the time required in order to transmit a *data packet*. The reservation subframe is divided into mini-slots, the length of each of which is equal to the time taken to transmit a *reservation request packet*. The mini-slots are grouped into pairs. Stations use random access (S-ALOHA) to contend for the mini-slots to make reservations on the uplink channel. Once they make it through, on the downlink channel an On-Board Scheduler (OBS) assigns slots in the information subframe for the stations to transmit data. If there is a collision of the reservation request packets, a binary tree contention resolution algorithm is employed to identify and schedule retransmissions of the collided reservation request packets. The pair of mini-slots in which the collision occurred is precluded from contention, while the other pairs are still open for contention. Then the binary tree contention resolution algorithm can be formulated as follows:

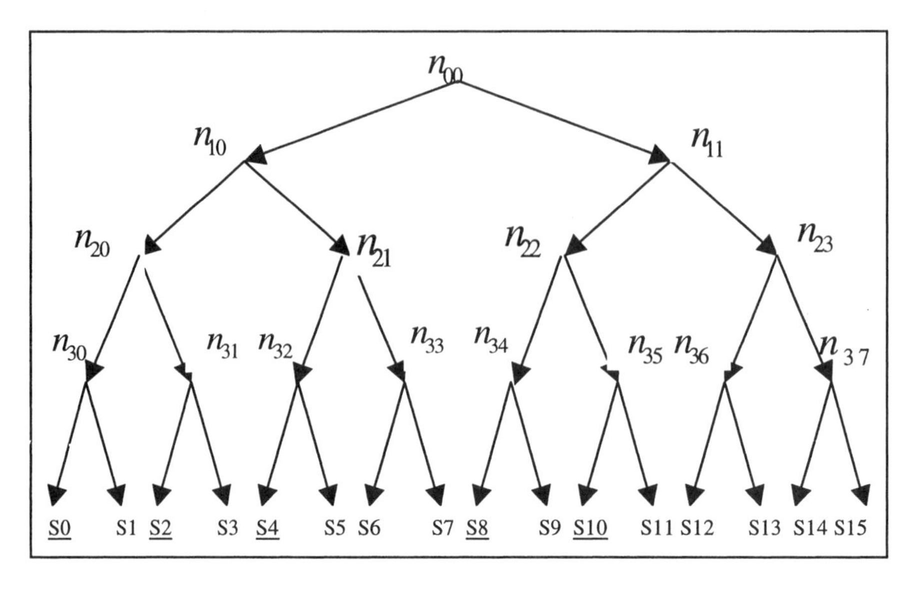


*Figure 3-15.* Stations as leaves of a binary tree

1) Let each of the stations be represented by a leaf of a binary tree [Capetanakis, 1979] as shown in *Figure 3-14*. All the nodes with the exception of the leaf nodes are "virtual" nodes. Each station has an associated binary address. Let $T_{x_1x_2}$ and $T_{y_1y_2}$ be two rooted sub-trees (the rooted sub-tree $T_{ij}$ is the tree with root node $n_{ij}$. In a binary rooted tree, $n_{ij}$ corresponds to the *j*th node at depth *i*), and assume that no collisions have occurred up to the beginning of the present pair of mini-slots.

2) Set $T_{x_1x_2} = T_{10}$ and $T_{y_1y_2} = T_{11}$.

3) If a station in $T_{x_1x_2}$ want to make a reservation, it transmits the reservation request packets in the first mini-slot of the any randomly chosen pair of min-slots, and similarly a station in $T_{y_1y_2}$ transmits its reservation request packets in the second mini-slot of a randomly chosen pair of mini-slots.

4) If any collisions occur in the preceding step, then no new reservation packets from the collided stations are transmitted until these collisions are resolved. Resolve the first collision (if any) before resolving the second (if any), and so on.

5) A collision in $T_{x_1x_2}$ (or $T_{y_1y_2}$) is resolved by dividing $T_{x_1x_2}$ (or $T_{y_1y_2}$) into two sub-trees (for example $T_{10}$ is now divided into $T_{20}$ and $T_{21}$) and then, repeating steps (3) and (4). Now the mini-slots in the same position are used in subsequent frames for continuing the resolution process. As mentioned earlier, stations other than the initial set of colliding stations are precluded from trying to access the pair of mini-slots in which the collisions occurred.

This algorithm is equivalent to the following tree search. Beginning with the root node and at each succeeding node, ask whether there are zero, one or more than one reservation request packets in the sub-tree stemming from each of the two emanating branches from that node. If the number of reservation requests generated at any of the sub-trees is more than one, then proceed to the node at the root of this sub-tree and repeat the question. This continues until all the leaves are separated into sets such that, each set contains at most one reservation packet.

For example, let there be 16 stations {S0, S1,......, S15}, and let each correspond to a leaf on the 16-leaf binary tree as shown in *Figure 3-15*. *Figure 3-16* depicts the slotted channel time. Note that the reservation mini-slots are paired. Now assume that no collisions have occurred until the beginning of Frame 1. It is assumed that the round-trip delay is less than one frame duration. Thus, any collisions in a particular frame are made known by the satellite on its downlink transmission before the beginning of the next frame. Beginning with Frame 1 where the first contention arises, the tree algorithm invokes the following steps in the indicated slot pairs.

Frame 1: Stations S0, S2 and S4 (all in $T_{10}$) suffer a collision of their reservation request packets in the first mini-slot of a pair; S8 and S10 in $T_{11}$, also suffer a collision in the second mini-slot of that pair. Then, any new reservation request packets from S0, S2, S4, S8 and S10 are precluded from transmission until the contention among the relevant stations is resolved. The particular pair of mini-slots in which the collision occurred is precluded from being open for further contention.

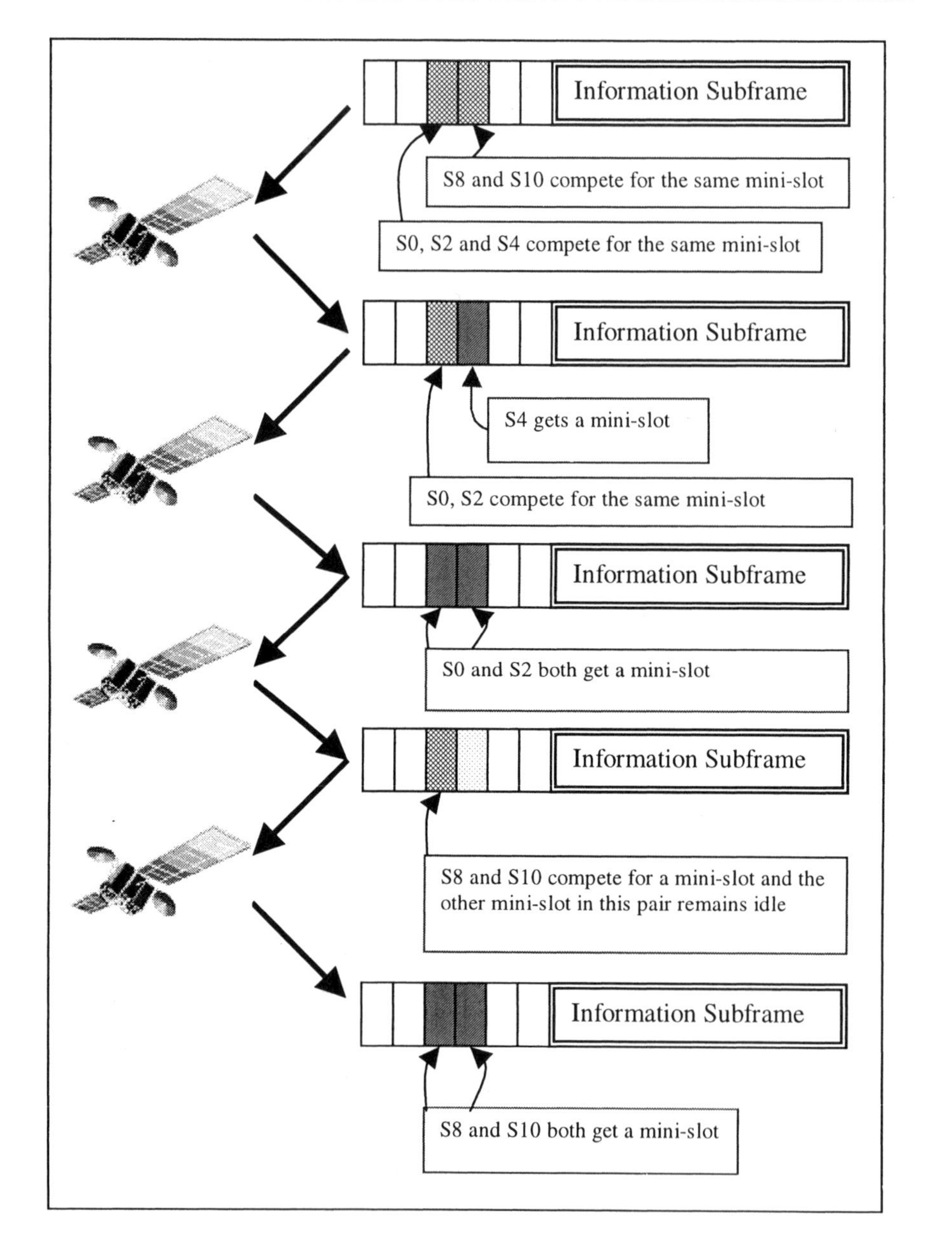


*Figure 3-16.* Example of the binary tree contention resolution algorithm

Frame 2: Since there is a collision in $T_{10}$, the stations in $T_{10}$ are divided in half and the previously collided reservation request packets in $T_{20}$ and $T_{21}$ are transmitted in the first and second slots of the pair of mini-slots (the same pair as in Frame 1), respectively. This results in a collision between the request packets of S0 and S2 and in a successful transmission of the transmission request packet of S4.

Frame 3: Since there was a collision in $T_{20}$, the collided reservation request packets in $T_{30}$ and $T_{31}$ are transmitted next in the first and second mini-slots of the same pair as before, respectively. This results in two successful transmissions by S0 and S2.

Frame 4: Since there was a collision in $T_{11}$, the collided reservation packets in $T_{22}$ and $T_{23}$ are transmitted in the successive frame using the same mini-slots. This results in a collision between the transmissions of S8 and S10 in the first mini-slot and no transmissions occur in the second mini-slot.

Frame 5: Since there was a collision in $T_{22}$, $T_{34}$ and $T_{35}$ transmit in the first and the second mini-slot of the pair, respectively. This results in successful transmissions by S8 and S10.

Frame 6: The original contention has been resolved. The pair of mini-slots under resolution is now open for contention again. The process continues, as described above.

The maximum throughput of CRA is 43% [Capetanakis, 1979]. Although promising, we see from the example that this protocol tends to be complex to implement.

### 5.2.5 Packet-Demand Assignment Multiple Access (PDAMA)

The Packet Demand Assignment Multiple Access is primarily designed to provide reservation-based access. The presence of an on-board scheduler (OBS) is assumed and this scheduler arbitrates channel access by processing the reservation requests. In PDAMA, time is once again divided into contiguous frames. The frame structure is shown in *Figure 3-17*. The frame is primarily divided into two parts: the information subframe and the control subframe.

The control subframe is further sub-divided into three parts: the leader subframe, the guard subframe and the request subframe. The request subframe contains several request slots (as shown in the *Figure 3-17*). These slots are open for contention access via the S-Aloha protocol. Transmissions by multiple users in a single request slot leads to a collision.

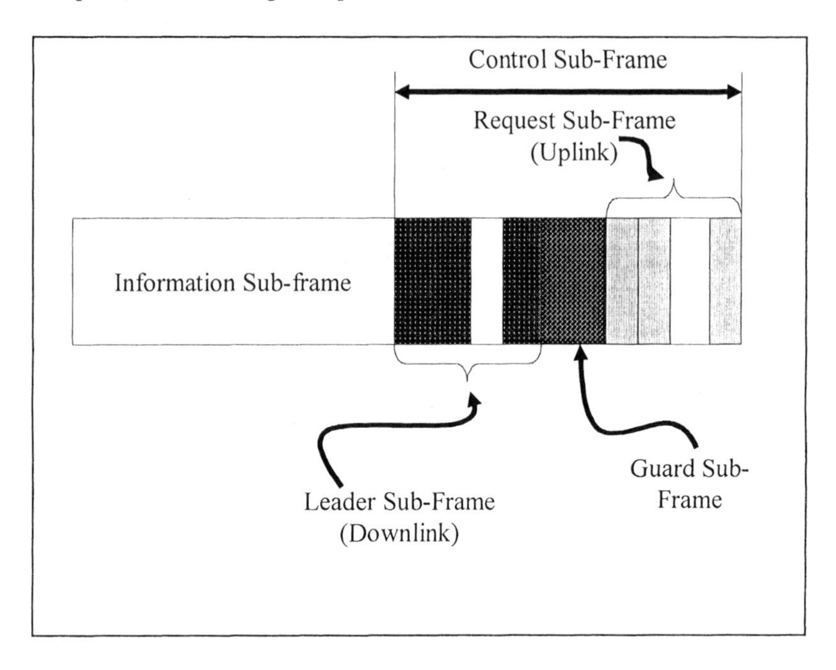


*Figure 3-17.* Frame Structure in PDAMA

The leader frame is used for downlink transmissions from the satellite to the earth stations. A global view of time is provided so that all the stations may synchronize themselves to a unified clock. Furthermore, in this subframe the satellite indicates the success of the reservation requests that were made in a prior frame (that occurred approximately one round trip time or 280 milliseconds earlier). In response to the successful reservation requests, slots are allocated in the information subframe. The downlink transmission in the leader subframe also indicates the scheduling information for the following information subframe. The scheduling is done by the OBS in accordance to a prioritized queue that it maintains. The information subframe is used for both uplink and downlink transmissions.

In PDAMA, it is recognized that the round-trip delay differs from station to station. Thus, there is the need for appropriately including guard bands to ensure time-synchronization among the stations. In order to separate the downlink transmissions in the leader subframe from the uplink transmissions in the request subframe a guard sub-frame is included. This subframe is typically 280 milliseconds long so that these subframes are sufficiently separated to accommodate all variations in the propagation delay. Furthermore, the scheduler on the OBS ensures that a station is not allocated transmitting and receiving slots consecutively in the information subframe to prevent the loss of data due to inaccuracies in the estimated propagation delay.

A station that transmits a reservation request monitors the leader frame to determine whether its transmission was successful. If it receives an acknowledgement within certain predetermined time (this time has to be larger than the round trip delay), it deems its transmission successful. If not, it retransmits the reservation request after invoking the exponential back-off algorithm [Pavey et al., 1986]. This back-off mechanism is essential to reduce the possibility of further, repeated, collisions.

# 6. HYBRID PROTOCOLS

Protocols that combine both reservation access and random access to improve performance are referred to as hybrid protocols. Such hybrid schemes have a built-in capability of adapt to the offered traffic profile. Thus, depending upon the intensity and the profile of the offered load, the protocols may dynamically adapt to offer opportunities for stations to either use random access or to make reservations.

## 6.1 Round-Robin Reservations (RRR)

RRR [Peyravi, 1999; Roberts, 1975] combines random access with reservations. In RRR it is assumed that there are more slots than stations, and each station can be assigned a home slot; the extra slots (those slots that are not home slots) are not assigned to anyone. If the owner of a home slot does not want it during the current frame, the slot goes idle. During the next frame, the slot becomes available to anyone who wants it, on a contention (ALOHA) basis. If the owner station wants to retrieve its home slot, it transmits a packet in that slot in the next possible frame, thus forcing a collision (if there was other traffic). After a collision, on the downlink the on-board arbiter broadcasts a message indicating that everyone except the owner station must desist from using the particular slot in the next frame. Thus, the owner station under discussion, in the worst case, can always regain a lost home slot within two frame times of when it might want to do so. At low channel utilizations, the system does not perform as well as S-ALOHA, since, after each collision in a slot, the slot must be closed for contention for at least one frame so that it may be determined if the owner station wants the slot back. Thus, if the owner station is not the station that caused a collision, the slot is simply wasted in the subsequent frame.

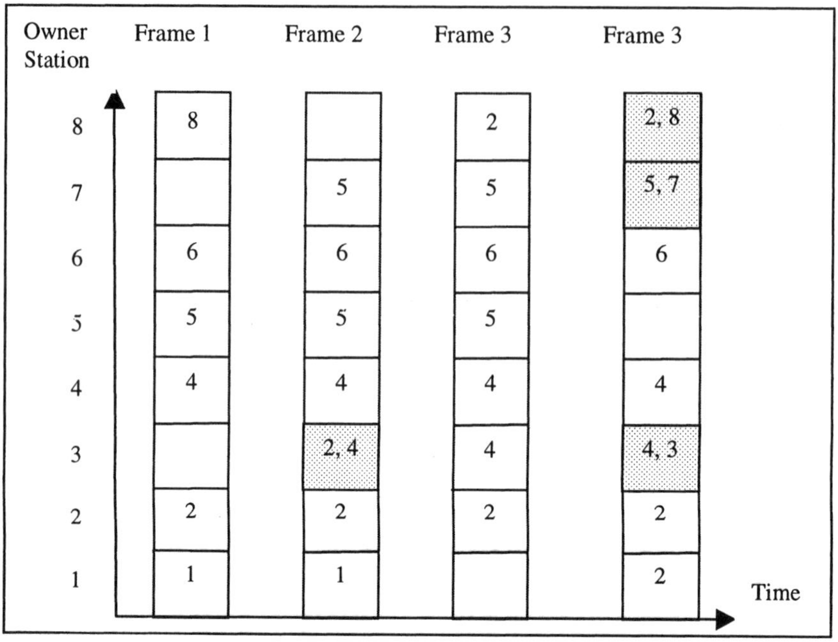

| Owner Station | Frame 1 | Frame 2 | Frame 3 | Frame 3 |
|---|---|---|---|---|
| 8 | 8 | | 2 | 2, 8 |
| 7 | | 5 | 5 | 5, 7 |
| 6 | 6 | 6 | 6 | 6 |
| 5 | 5 | 5 | 5 | |
| 4 | 4 | 4 | 4 | 4 |
| 3 | | 2, 4 | 4 | 4, 3 |
| 2 | 2 | 2 | 2 | 2 |
| 1 | 1 | 1 | | 2 |

Time

*Figure 3-18.* The mechanics of the RRR protocol

In a variation of this scheme, each station keeps track of "a virtual global queue". The maintenance of this queue is facilitated by each station by including the length of its own local queue (containing the packets generated by that station) in the header of every packet that it sends. The information is then relayed on the downlink by the satellite. A round-robin algorithm is invoked to access available slots (excess slots and unused slots) in accordance with the scheduling rules of this virtual queue. RRR can accommodate bursty data traffic and is adaptive to variations in station population. This approach is superior to R-Aloha for stream-dominated traffic since each station is guaranteed a slot per frame when it has traffic to send. When there are a large number of stations, the use of RRR can lead to large delays. Furthermore, as discussed earlier, the methods used to open unused slots for contention and to retrieve these slots for fixed access when the need arises can result in a wastage of bandwidth in some scenarios.

In *Figure 3-18* we consider an example to illustrate how RRR functions. In frame 1, the slots owned by stations 3 and 7 are idle and thus available for acquisition. In frame 2, station 5 successfully acquires the slot owned by station 7 while the transmissions of stations 2 and 4 collide in the slot owned by station 3. In addition, the slot owned by station 8 is now idle. Frame 3 shows stations 2 and 4 successfully acquiring the slots owned by stations 3 and 8, respectively. Finally, frame 4 depicts stations 3, 7 and 8 causing collisions in their home slots to indicate that they want to get their slots back.

## 6.2 Interleaved Frame Flush-Out (IFFO)

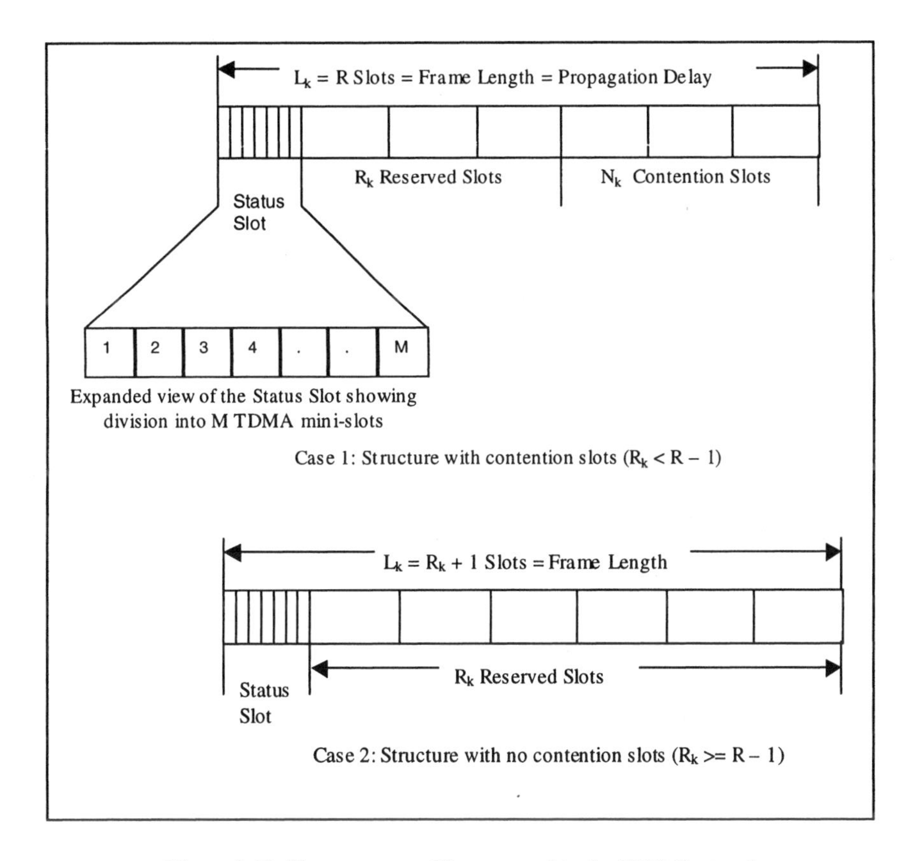


*Figure 3-19.* The structure of frames used in the IFFO Protocol

The Interleaved Frame Flush-Out (IFFO) protocols [Wieselthier and Ephremides, 1980] incorporate both contention and reservation based access for arbitrating the use of the satellite channel. These protocols are especially suited for use in packet switched satellite networks. They have a built in capacity to adapt to variations in traffic levels and profiles. The frame structure used in IFFO protocols is shown in *Figure 3-19*. A frame k contains a set of reserved slots, $R_k$, and a set of unreserved slots $N_k$. The unreserved slots may be used for transmission on a contention basis. The first slot of each frame is called the status slot. It consists of M mini-slots that are assigned to M stations in accordance with TDMA (contention-free). Stations use these M mini-slots to reserve bandwidth for transmitting data packets. Therefore, M, the number of reservation mini-slots that can be filled in "one slot duration" limits the number of stations that can be admitted in the network. The position of the mini-slot in which a station transmits establishes the identity of the transmitting station. Thus, the reservation mini-packets do not have to carry the identification of the sender or any synchronization information, and hence, they can be fairly small in size.

The status slots are followed by $R_k$ reserved slots. It is required that each frame length contains at least a total of R slots (R is the round trip delay in slots), in order to assure that the reservation information generated at the beginning of each frame is received by all stations (through a broadcast by the satellite) before the beginning of the next frame. Thus, the reservations made in a frame are typically honoured in the immediately following frame. The length of the frame ($L_k$) can be considered to be a sum of the status (one), the reserved ($R_k$) and the unreserved ($N_k$) slots. When the number of reserved slots $R_k$, in a frame k, is less than R, the remaining $N_k$ slots are used for transmission on a contention basis (see *Figure 3-19*: case 1), by using the S-ALOHA Protocol. The frame structure and length, therefore, adapt to channel traffic.

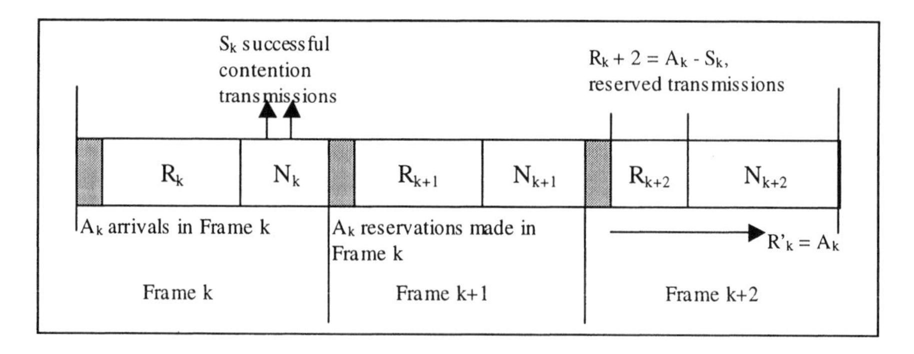


*Figure 3-20.* IFFO reservation scheme

The boundary between the reserved slots and the contention slots is *movable* and depends on the number of reservations made in the previous frame. Thus, if the all slots in the frame are reserved, then, there are no contention slots (*Figure 3-19*: case 2). Let us consider a frame, 'k' in which packets are being transmitted. If a packet arrives during frame k, a *Figure 3-20*). However, there is a possibility that there may be contention slots available during frame k itself (*Figure 3-19*:case 1). In such a case, the newly arrived packet can be transmitted in one of these contention slots. If a collision occurs in the contention slot then the packet is retransmitted by making an explicit reservation request and using the allocated slot in response to the request. We reiterate that while the reservation is made in frame k+1, the packet is transmitted in frame k+2 (as shown in *Figure 3-20*) when the reserved slot is actually obtained. Thus, for a packet that arrived in frame k, if an attempt to transmit the packet in a contention slot in frame k fails, the packet is transmitted with a reservation in frame k+2. The information with regards to the successfully transmitted packets in frame k is made available to all stations by means of a downlink (broadcast) transmission. This control information for any packet in frame k is made available by the end of first slot (status slot) of the frame k+2. If the packet transmission in a contention slot is successful, then the reservation made for the packet in the next frame (frame k+1) is cancelled. This cancellation is achieved by observing the information sent with regards to successful packet transmissions ($S_k$) via the contention slots ($N_k$) by various stations, on the downlink broadcast. The stations then, accordingly, delete their corresponding reservations $S_k$. This is shown in the *Figure 3-20*, where the number of reservation transmissions in frame k+2 is reduced from $A_k$ to $A_k$ - $S_k$ ($A_k$ is the number of packet arrivals in frame k, and $S_k$ is the number of successful transmissions using the contention slots in frame k). Importantly, if a packet is transmitted in a contention slot in frame k, then there are no attempts made to transmit the packet in the $(k+1)^{st}$ frame. If the attempt to transmit a packet in the contention slot in frame k fails, then the packet is transmitted in the reserved slot in the $(k+2)^{nd}$ frame. Note that the slot reservation process in any 'even' numbered frame is independent of the slot reservation process in an odd numbered frame. Thus, any packet that arrives in a particular frame k is "flushed out" typically by the end of frame k+2. This motivates the name "Interleaved Frame Flush-out".

The protocol may be varied by incorporating a different transmission procedure for accessing the unreserved slots. In the *Pure Reservation IFFO* scheme, the stations refrain from transmitting in unreserved slots. In the *Fixed contention IFFO* scheme, a new packet is transmitted in a contention slot only if it arrived during the period between the slots ($R_k$+1) and (R–1). In the *Controlled Contention IFFO* scheme, a station transmits a packet in an unreserved slot with a certain probability which is a function of the number of packets that were generated in frame k, the slot position at which the packet arrived within the frame and number of reserved slots in the frame ($R_k$).

## 6.3 Split-Channel Reservation Upon Collision (SRUC)

The Split-Channel Reservation Upon Collision (SRUC) protocol [Borgonovo and Fratta, 1978] is another adaptive hybrid scheme, where the media is primarily accessed through contention. A superimposed reservation scheme is invoked whenever a collision occurs. While primarily suitable for multiple access on links with large propagation delays, SRUC can also be used in systems in which propagation delays are small. SRUC divides the stations (M) that are competing for channel access into a number of groups (F), each consisting of N stations (i.e., F = M/N). Time is divided into contiguous frames, each of which contains F slots (see *Figure 3-21*). Each group is assigned one particular slot in the frame. Each slot is divided into two sub-slots, Sub Slot 0 (SS0) that is used for the transmission of information (data and header) and Sub Slot 1 (SS1) that is used for the transmission of signalling information (SI). Each individual SS1 is further subdivided into N parts (N is the number of stations that form a group). Signalling information is transmitted without any contention by each of the N stations of a group in Sub Slot 1 (SS1) of the slot that belongs to that group. The size of the sub-slot, SS1, is relatively small since each of these slots serves only one group of stations (of size N << M).

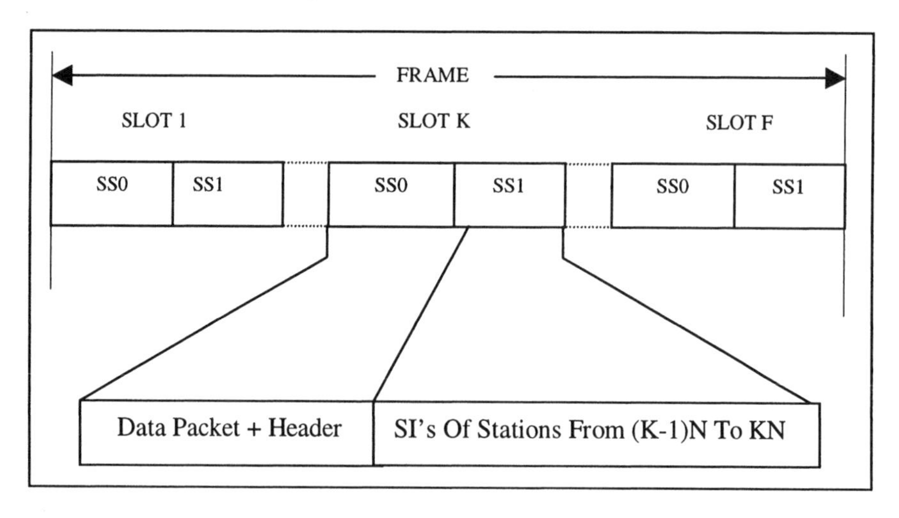


*Figure 3-21.* SRUC channel subdivision

In SRUC each station is required to maintain a queue that is called the Request Queue (RQ). When a station has a packet to transmit, it transmits a reservation request for that packet in the part of the SS1 slot that belongs to its group and is assigned to it. The satellite (a bent pipe satellite is assumed) broadcasts the information with regards to this reservation request on the downlink. The request is then inserted into the RQ by all the stations within the satellite footprint. Thus, the queue RQ, at each station, contains global information with regards to all the requests that are made for channel access. If multiple requests are indicated simultaneously on the downlink, these requests are ordered in the RQ in accordance to predefined priorities that are associated with the requests. The channel access is controlled by a protocol called the Data Channel Access Protocol or DCAP. The functionalities of DCAP are described in the following paragraph.

The owner station that has its packet at the head of the RQ will access the immediately subsequent SS0 slot to transmit the packet associated with the request. The packet is then removed from the RQ. When the RQ becomes empty, the data slot, SS0 is now open for contention access. When a station has a new packet, it transmits the packet using the contention slot. In spite of the fact that an attempt is made to transmit the packet via a contention slot, the owner station also makes an explicit reservation request via its SS1 slot so that, in the wake of a collision a reserved slot may be obtained at a later time. The status of a transmission in the SS0 slot is also made known on the downlink. Thus, if a packet is successful via contention, the reservation request for the packet is not inserted into the RQ.

The above channel control procedure (CCP) can be implemented in a distributed or centralized fashion. Since all stations can hear the downlink retransmissions from the satellite, a distributed implementation of channel control procedure, as described in this section, is possible. The SRUC protocol is always stable since all colliding packets are retransmitted by using reservations. Note that the sub-slots for signalling can be implemented as separate frequency channels instead of separate time channels.

## 6.4 Announced retransmission Random Access (ARRA)

The Announced Retransmission Random Access Scheme (ARRA) [Raychaudhuri, 1985] is a hybrid MAC protocol that uses contention-based access as the primary mode for accessing the channel. ARRA achieves a higher utilization as compared to traditional random access methods by using the satellite downlink to disseminate control information that helps in considerably reducing the possibility of collisions. Each station transmits explicit control information on a separate uplink channel to the satellite. The satellite acts like a bent pipe. It receives the messages from all of the stations within its footprint and broadcasts the received messages on the downlink.

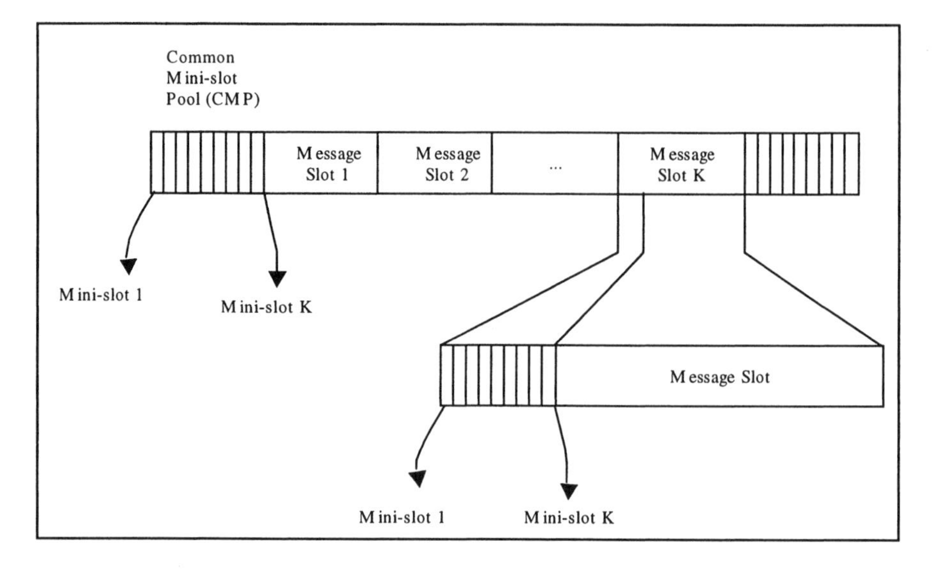


*Figure 3-22.* Frame format for ARRA

In ARRA a frame is divided into contiguous time slots as shown in *Figure 3-22.* Each of these time slots is capable of accommodating a complete MAC layer message. In each slot, the allocated bandwidth may be further sub-divided into two parts. The first part consists of a predetermined number of mini-slots (say K) and is K bits in length. In other words, each mini-slot is exactly one bit long. The second part of the slot is used for the actual transmission of the message. When a slot (while we use the term "slot" here, what we actually mean is the second part of the slot that is used for message transmission) is available for contention access, a station might choose to transmit in that slot. However, it anticipates the possibility that the transmitted packet might collide (since other stations might also choose to transmit in the same slot). Thus, it indicates intent to transmit again, if necessary, in an arbitrarily chosen slot (say the $i^{th}$ slot) of the subsequent frame by sending a message in the $i^{th}$ mini-slot of the first part of the slot.

The stations upon receiving a broadcast from the satellite examine the set of mini-slots for that slot. If nothing was transmitted in this set of mini-slots, then this indicates that no station attempted to access that particular slot in the uplink. On the other hand, if there were multiple mini-slots in which transmissions were received, it is evident that multiple stations attempted transmissions in that slot and as a result there was a collision. The stations *aggregate* the information contained in all the mini-slots of the frame and infer as to which of the slots in the following frame, have been chosen for retransmission attempts of collided packets. This set of slots is now exclusively *reserved* for retransmissions and newly generated packets are precluded from being transmitted in those slots. Thus, in order to transmit a newly generated packet, the owner station will have to choose a slot from the remaining slots in the frame.

If all the slots in a particular frame were claimed for performing re-transmissions, a newly generated message is declined access. ARRA provides a mechanism for this newly generated message to make a reservation in a subsequent frame. The frame is modified to contain a separate pool of mini-slots. This pool is not affiliated with any of the message slots and is called the Common Mini-Slot Pool or CMP. The owner station of the newly generated packet, upon being denied access in a frame, will indicate the intent to transmit in a particular slot (say the $j^{th}$ slot) of the subsequent frame by sending a message in the $j^{th}$ mini-slot of the CMP. The contents of the CMP are aggregated along with the contents of the other mini-slots in the frame. Thus, the newly generated packet receives a reserved slot in a subsequent frame.

One possible way of performing the aggregation of the received "intent to retransmit messages" is to simply exclude the slots indicated in the mini-slots of the slots where collision occurred from being open for access to newly generated packets. This version is called the "basic version" of ARRA. However, in many cases, two stations might choose the same mini-slot (say two stations that failed in attempts to transmit in separate slots in a given frame, both indicate an intent to transmit in the $k^{th}$ slot of the frame following the downlink broadcast from the satellite). In such a case, if such retransmissions were to be allowed it would result in another collision. This may be avoided by an indication by the satellite that there was a "collision" in the particular mini-slot under discussion. The relevant stations would then refrain from transmitting in the corresponding slot (in our example, this corresponds to the $k^{th}$ slot). The slot is now open for contention to be used by stations that have newly generated messages. The collided messages will now attempt to reserve slots in a subsequent frame by using the mini-slots in the CMP. This version of ARRA is called the "extended version".

ARRA retains the simplicity and the operational convenience offered by conventional random access protocols such as S-Aloha. In addition, it offers a throughput of about 50% (basic ARRA) to about 60% (extended ARRA). Note that this is considerably higher than that offered by a conventional random access protocol such as S-Aloha (37%).

## 6.5 Scheduled-Retransmission Multiple Access (SRMA)

The Scheduled-Retransmission Multiple Access or SRMA [Yum and Wong, 1989] efficiently combines both random access and reservation access methods. Two versions of SRMA are described in [Yum and Wong,

1989]: the fixed-frame version (SRMA/FF), and the dynamic frame version (SRMA/DF).

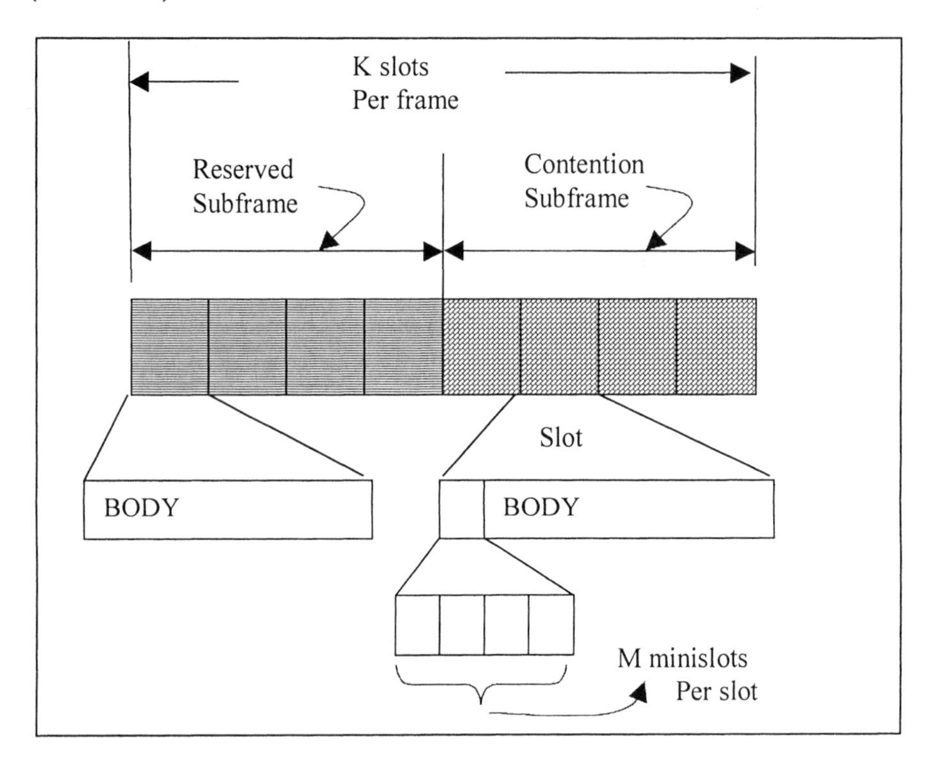


*Figure 3-23.* Frame structure of SRMA protocol

In the FF version, time is divided into frames of K slots each. Each frame is divided into a contention subframe and a reserved subframe (*Figure 3-23*). Each slot in the contention subframe has a header field and a body field. The header consists of M minislots. The body field of the slot can accommodate one information packet. Note that guard times are needed between frames, slots, and minislots to assure synchronization. New packets that arrived during the contention subframes are transmitted immediately and for access control, the S-ALOHA scheme is employed. The packets that arrive during the reserved subframes, however, are scheduled to be retransmitted in one of the 'U' (U is a system parameter that can be configured) upcoming contention slots at random. If an attempt to transmit a packet in a contention slot is made, one of the M minislots in the corresponding header of that slot is marked at random for the purposes of scheduling a retransmission in the case of a collision in that slot. Let the round-trip propagation delay be R frames. It takes R frames after the initial transmission for a station to learn via a feedback channel (a low-rate announcement subchannel from the satellite to the stations), whether the transmission was successful or not. The station with an unsuccessful transmission will be assigned a reserved slot in the next frame for the retransmission of the packet if the retransmission reservation request, made in the mini-slots, is successful. The collection of all the reserved slots (a consequence of all such reservations) in a frame constitutes the reserved subframe. The packets retransmitted in the reserved subframes do not encounter collision. If both the transmission and the retransmission reservation request are unsuccessful, the owner station will attempt to retransmit the packet in one of the 'V' (V is also a configurable system parameter) upcoming contention slots chosen at random.

Each packet is assigned a status vector represented by the ordered pair, (x, y). Here $x \in \{1, 2, \ldots, K\}$ is the slot position in which it is attempted to transmit the packet and $y \in \{1, 2, \ldots, M\}$ is the position of the minislot in which the owner station of the packet indicates an intent to retransmit if the packet were to collide. When a frame is received by the satellite, the on-board scheduler collects the set of such vectors and processes them in accordance with the following procedures:

a) All packets that were transmitted simultaneously in a particular slot, and hence, have collided with each other, but have different status vectors are scheduled for later retransmission. (The (x, y) values are used for scheduling the transmissions. Two packets with the same (x, y) value cannot be differentiated and therefore, cannot be rescheduled for transmission.)
b) The rescheduled packets are sorted in an increasing order with respect to x. The set can now be divided into subsets, the elements of each of which have the same value of x.
c) The elements of each subset are now arranged in an increasing order with respect to y.
d) The resulting set of ordered status vectors is truncated to a size of K, since, in the retransmission frame at most K reservations can be made. Those packets that do not belong to the truncated set will have to wait for the next retransmission frame in order to receive reserved slots for retransmissions.

This set of ordered status vectors is then broadcast to all stations by the satellite together with acknowledgements to successfully received packets. A station that is anticipating an acknowledgement but does not receive one, will search for its status vector. If its status vector indicates an allocated slot say, the $b^{th}$ slot, the station will retransmit its packet in the $b^{th}$ slot of the next frame. If, on the other hand, the status vector is not found, the reservation request for performing the retransmission is deemed unsuccessful. The station then contends for the channel after a random delay.

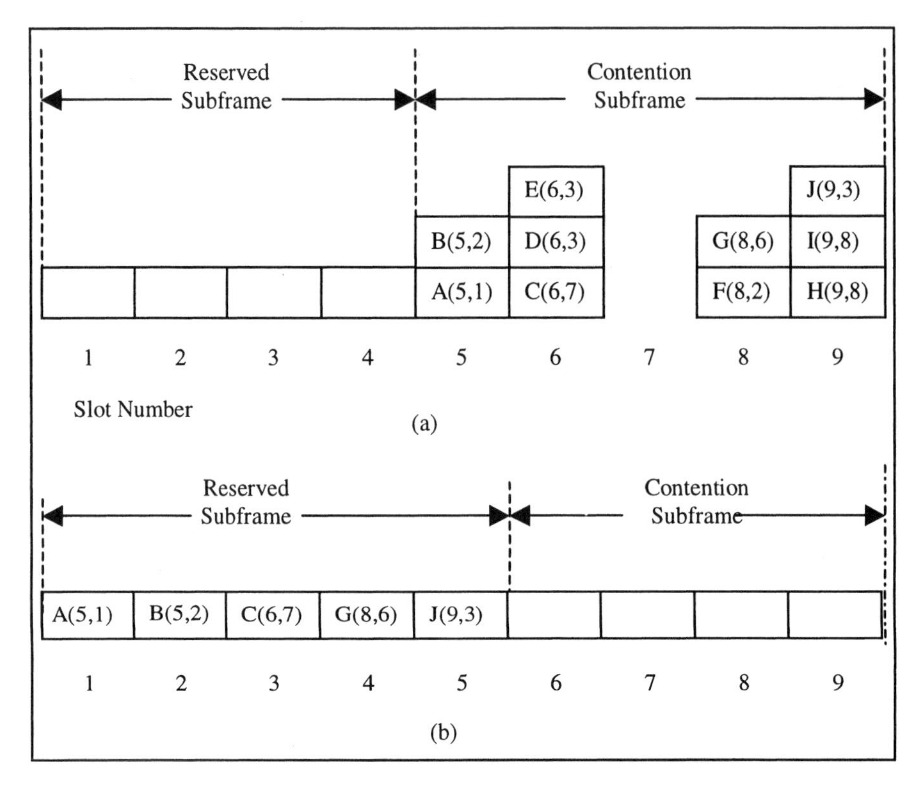


*Figure 3-24.* Retransmission reservation for SRMA/FF protocol

To illustrate the functioning of this protocol, let there be ten slots in each frame and eight minislots in the header of each slot (i.e., K = 10, M = 8). A typical frame is shown in *Figure 3-24*(a). Here, the contention subframe starts at slot 5. Two packets, A and B collide in slot 5. As shown, the owner stations choose the corresponding minislots positions 1 and 2 respectively, for requesting reservations for retransmissions. In slot 6, three packets collide. Slot 7 is empty. Two packets with status vectors (8, 2) and (8, 6) collide in slot 8, and so on. In accordance with policy (1), packets D, E, I and H are to be discarded. As per policies (2) and (3) the packet status vectors are arranged as (5, 1), (5, 2), (6, 7), (8, 2), (8, 6) and (9, 3). This is the order in which these packets will be retransmitted in the reserved subframe, R frames later. *Figure 3-24*(b) shows the corresponding retransmission frame.

In the dynamic frame or DF version, the framing pattern is different. In this version a contention frame (not a subframe) is present, and has a fixed length of F slots. There is also a reserved frame that has a length equal to the number of successfully reserved packets in its previous corresponding contention frame (R frames earlier). Thus, in contrast with the FF version, there is no limit on the number of reserved slots in a frame in this version. Using the DF version, a station having a packet to transmit would 1) transmit immediately if the channel is in the contention mode; or 2) if the channel is in the reserved mode, wait for the beginning of the next contention frame, and transmit the packet in one the F slots chosen at random in that contention frame. The feedback information that is transmitted by the satellite on the downlink is similar to that in the FF version except that now, the status of all successfully scheduled packets is sent back right away (no truncation). The DF version requires all stations to keep track of the status of all reserved frames in terms of when they occur and for how long they last. This is obviously essential for frame synchronization purposes. In contrast, the FF version requires the stations to keep the length of the current reserved subframe only.

SRMA/FF achieves a maximum throughput of 65%, while SRMA/DF can achieve a maximum throughput of 89%. Both versions of SRMA can help achieve average delays considerably lower than that achievable using S-ALOHA, at high loads. In summary, SRMA helps achieve the low latency of the ALOHA protocols at low loads while emulating reservation access at high loads.

## 6.6 Response Initiated Multiple Access (RIMA)

RIMA [Connors and Pottie, 2000] considers the problem of optimizing the media access wherein the last hop of Internet access is through a satellite link. Such a system consists of a number of earth stations sending and receiving packets from a gateway terminal that communicates directly with the satellite. The optimization of the Media Access Control (MAC) is possible since the channel requirements for stations can be predicted with a reasonable accuracy when bulk of the traffic is generated from applications that use Transmission Control Protocol / Internet Protocol (TCP/IP) suite of protocols. Bulk of the traffic carried in the Internet is by using TCP/IP. Many applications, such as HTTP (used for surfing the web), FTP (File Transfer Protocol), etc use TCP/IP. When the applications use TCP/IP, the size of the packets received at the MAC layer are either approximately equal in size to the Maximum Transmission Unit (MTU) size (these packets are referred to as data packets) of the MAC protocol in use, or are fairly small in size (these are acknowledgement packets). Thus, the use of TCP/IP leads to a bimodal distribution of packet lengths at the MAC layer. Furthermore, the transmission of a TCP/IP data packet can be expected to entail an acknowledgement packet in response (with high probability) and vice versa. As we see later in this section, this fact can be used in reserving bandwidth on the uplink channel.

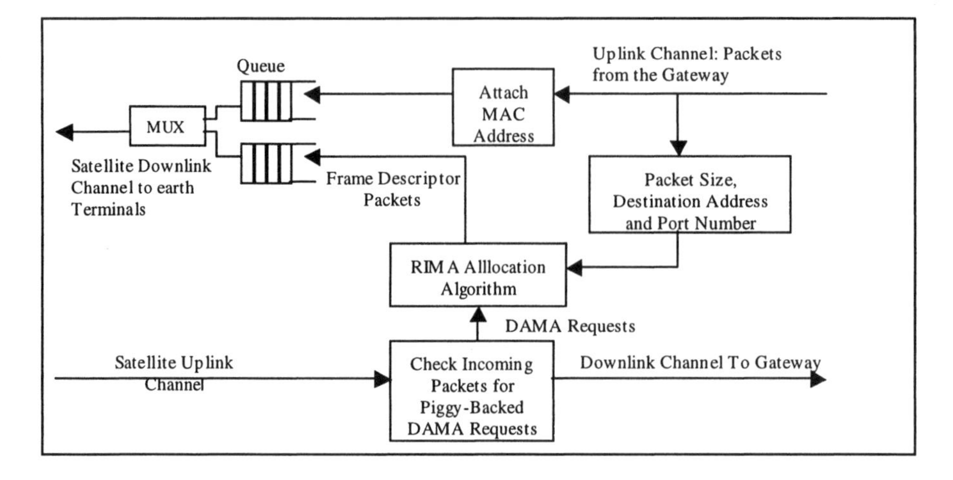


*Figure 3-25.* Functional diagram of RIMA

RIMA uses a framework wherein, time is divided into contiguous frames which are further subdivided into slots (TDMA). At the beginning of each downlink Time Division Multiplexed frame, the RIMA allocating agent (AA) located at the satellite issues a *frame descriptor packet* (FDP). This packet specifies the slot assignments in the frame on the uplink channel. Unassigned slots are available for Random Access to all stations. The FDP is generated by using two inputs: the RIMA allocation algorithm (to be described) operating on the packets arriving via the gateway and additional bandwidth requests received from the earth stations (see *Figure 3-25*).

The RIMA allocation algorithm may be described as follows. Consider a packet that was sent by the gateway and received at the satellite. The allocating agent at the satellite uses the destination IP address in the header of the packet to determine the earth station to which the packet is being sent. Furthermore, it uses the port number (also found in the header of the header of the packet) and the size of the packet to estimate the uplink bandwidth requirements of the predicted response from the packet's destination (an earth station). For example, if the AA determines the port number in the header of the packet to be equal to 21 (ftp application) and the size of the received packet to be 40 bytes (size of a TCP acknowledgement packet), then it assumes that, in response to this acknowledgement packet, the destination station will generate an "MTU sized" data packet. Similarly, the AA can estimate the uplink bandwidth that might be required by various earth stations in order to respond to a large packet (i.e., a TCP data packet that will generate an acknowledgement packet). The uplink channel bandwidth allocation mechanism implemented by the allocating agent also responds to explicit bandwidth requests received from the earth stations. These explicit bandwidth requests can be "piggy backed" on the data packets as in a DAMA (see section 2.4) like scheme. If the bandwidth requirements of the stations are estimated accurately and the slots are accordingly assigned in the uplink frame, then the number of stations competing for random access is reduced. This has an effect of reducing the collision space of the uplink channel. Under the assumption that the variation in the round trip time of a packet is small, RIMA is shown to consistently outperform R-Aloha by 30 % [Connors and Pottie, 2000].

## 6.7 Combined Free/Demand Assignment Multiple Access (CFDAMA)

The Combined Free/Demand Assignment Multiple Access (CFDAMA) scheme is suitable for broadband packet satellite systems that serve a finite number of stations [Le-Ngoc and Krithnamurthy, 1996; Le-Ngoc and Jahangir, 1998; Krishnamurthy, 1994]. CFDAMA schemes are primarily based on the allocation of bandwidth to individual stations via reservations. However, when there are no reservations, the arbiter, (which might either be on-board or on-the-ground) assigns channel capacity in accordance to what is called *free* assignment. This *free* assignment may be made in accordance with some pre-determined strategy. A simple strategy would be to assign bandwidth to users based on a round-robin scheme. By combining the two strategies of free and demand assignment, CFDAMA schemes achieve short delays at low and medium traffic loads while maintaining the high channel utilization efficiency of DAMA schemes at high loads. While primarily used for the transmission of bursty data traffic, extensions to CFDAMA to allow the incorporation of real-time traffic (such as voice) is possible [Krishnamurthy, 1994].

In CFDAMA, time is divided into contiguous frames. If there are 'N' stations that are accessing the uplink using CFDAMA, then each frame facilitates the transmission of 'N' data packets. In addition, in each frame, there is a set of mini-slots that is used for making reservations. Each data-slot (used for the transmission of data packets) may have an associated mini-slot, or all the mini-slots may be grouped at the beginning of the frame. CFDAMA allows three possible strategies in order to make reservations. The aforementioned reservation mini-slots may be pre-assigned to the individual stations. This is usually done such that the time duration between consecutive assignments of request slots to a particular station is exactly equal to 'N' slots. This version of CFDAMA is called CFDAMA-PA, where PA is used to signify pre-assignment. In contrast, the reservation mini-slots may be open for contention or random access. This version of CFDAMA is called CFDAMA-RA (RA stands for Random Access). A third way of making reservations would be to piggyback reservation requests onto actual data. This version of CFDMA is called CFDAMA-PB, where PB signifies piggybacking. Note that for CFDAMA-PB, it is no longer necessary to have a separate set of mini-slots to allow reservation requests.

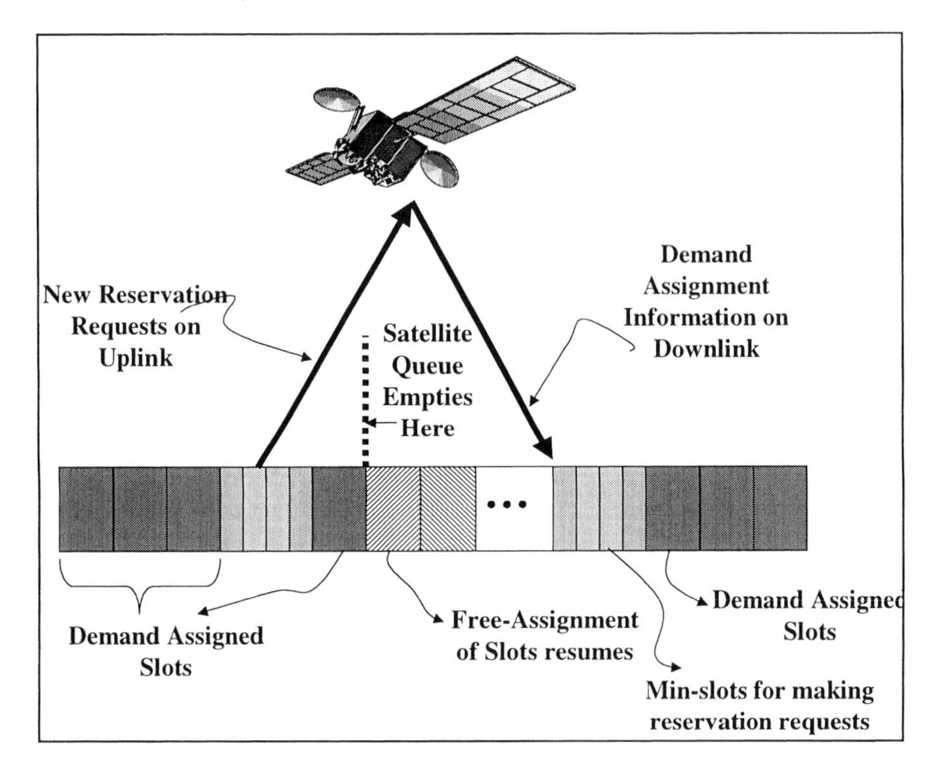


*Figure 3-26.* CFDAMA operations

The mechanisms of CFDMA may be illustrated with the help of *Figure 3-26*. In this figure we assume an on-board processor but the mechanics are similar when we instead, have an on-the-ground arbiter. When the satellite scheduler queue is non-empty, i.e., there are pending reservation requests, the slots are allocated in response to the requests. In the absence of requests (note that once a request is transmitted on the uplink it takes about 270 milliseconds before the satellite can recognize the request), the satellite allocates slots in accordance with the free assignment strategy. Once it receives further requests, it resumes the demand assignment strategy. In the figure, notice that in spite of the fact that there are new reservation requests sent on the uplink, they are not honoured until a round-trip delay later.

For low and medium sizes of station population, the CFDAMA-PA scheme is seen to have an excellent delay throughput performance. However, as the number of stations increases, the performance of

CFDAMA-PA degrades due to the long waiting time that a station has to endure before it gets its pre-assigned slot. The performance of CFDAMA-PB is very similar to that of CFDAMA-PA. However, it is possible for a few stations to dominate channel access by making repeated piggybacked reservations if CFDAMA-PB is used. Thus, fairness among stations may be compromised. CFDAMA-RA is very efficient if there are a large number of lightly loaded stations. However, the performance of the scheme degrades with an increase in load due to an increased number of collisions in the request mini-slots. A CFDAMA scheme that not only allows the piggybacking of reservation requests but also contains a small set of reservation mini-slots that are open for contention access can provide the best performance over a wide range of population sizes and varying traffic intensities.

## 6.8 Fixed Boundary Integrated Access Scheme (FBIA)

The fixed boundary integrated access (FBIA) scheme [Suda et al., 1983] integrates the use of S-ALOHA and reservations on demand and can handle different kinds of traffic including bursty traffic in short spurts and sustained heavy loads of data. The satellite is assumed to have an on-board processor that examines reservation request packets transmitted on the uplink channel and creates a downlink control packet to reflect these reservations. It is assumed that there are two kinds of stations, namely, stations that generate short spurts of bursty data traffic and stations which produce sustained heavy loads of stream traffic. The channel time is divided into frames, each of which is further divided into two subframes. The frame structure is shown in *Figure 3-27.* Note that each frame consists of time slots, the duration of each of which is equal to the time taken to transmit one data packet. It is assumed that all the stations are synchronized with each other and that each station can transmit a packet only at the beginning of a slot. One of these subframes, called the contention subframe, is used for accommodating bursty stations. The primary mode of access is contention-based access and the S-ALOHA protocol is employed. The contention subframe consists of a constant (K) number of successive slots. The other subframe, which is referred to as the reservation subframe, is used for accommodating stations with sustained heavy loads of traffic. This subframe is further divided into two mini-frames, a reservation mini-frame and a message mini-frame. These mini-frames contain a constant number of slots, say, V and S slots respectively. The reservation mini-frame is for transmitting reservation requests using the S-ALOHA protocol. These reservation requests facilitate the reservation of data slots in an upcoming message mini-frame. The V slots that are in the reservation mini-frame are further divided into L minislots, the duration of each of which is equal to the transmission time of a reservation request packet.

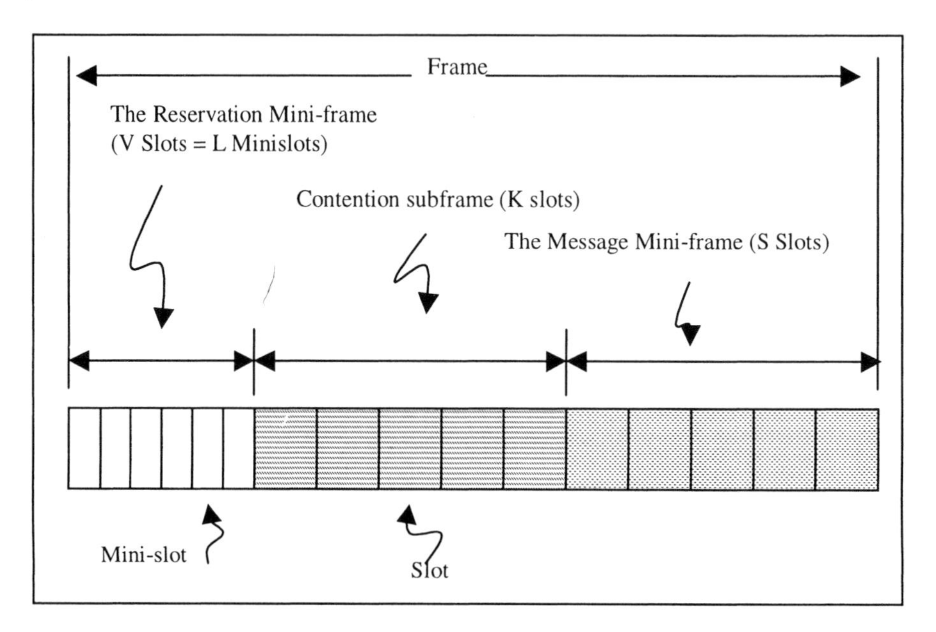


*Figure 3-27.* Frame structure used in the fixed boundary integrated access scheme

A bursty station having a packet to transmit at the beginning of the contention subframe chooses a slot from the set of slots in this subframe randomly, in accordance to a uniform distribution, and transmits its packet in the selected slot. After a round-trip propagation delay of R slots, the station will know (thanks to the control information on the downlink) whether the transmitted packet has been successfully received at satellite. If the packet has collided with another simultaneous transmission, it will be retransmitted in the succeeding contention subframe in accordance with the same rules as before.

A contention-based *reservation* scheme is employed by stations when they have large messages or sustained heavy loads of traffic to transmit. First, a station which has generated such traffic divides each message into fixed length packets, the size of each of which is the same as that of a data packet arriving at a bursty station. The station, then, creates a reservation request packet containing its identification. It transmits the reservation packet in a randomly selected mini-slot in the reservation mini-frame. The reservation is for acquiring slots in the message mini-frame for the purpose of transmitting the packets from the generated message. Upon receiving the reservation request packet successfully, i.e., without collision, the satellite attempts to assign a slot in the message mini-frame for the station. If there are any unreserved slots in the following message mini-frame, the satellite randomly assigns one of them to the owner station of the reservation request packet. If all of the slots are reserved, the reservation request is rejected. The station is informed of the success or failure of the reservation request by means of the control packet generated and transmitted on the downlink, by the satellite.

Once the reservation is successful, a slot in the exact same position is reserved in each of the forthcoming message mini-frames in subsequent frames for the station until the satellite receives an end-of-use signal from the station (implicit reservation). This end-of-use signal informs the satellite that the slot may be released for other use. If K (the number of slots in the contention subframe) is assumed to be greater than or equal to a round trip propagation delay R (in slots), the station will know whether a request for the reservation of slots is successful or not before the beginning of the message mini-frame in the frame. In case the reservation request packet collides due to uplink contention, or if a reservation failure occurs at the satellite, the station transmits the reservation packet in a randomly selected minislot in the next reservation mini-frame. In the case wherein $K < R$, the station cannot know as to whether there were slots assigned and if so, where they lie within the message mini-frame before the beginning of the immediately following message mini-frame. Thus, if the request was successful the reservation will be valid in a subsequent frame.

In this scheme there is a "*fixed boundary*" between the message mini-frame and the contention subframe. Thus, stations that are classified as being heavily loaded cannot transmit in the latter if there is unused capacity in that subframe. Similarly, those stations that are classified as bursty stations cannot transmit in the message mini-frame, even if many of the slots in that mini-frame are not used. This "hard" division of the channel capacity is not flexible to efficiently support dynamic traffic patterns.

## 6.9 Combined Random/Reservation Multiple Access (CRRMA)

The Combined random/reservation multiple access (CRRMA) protocol [Lee and Mark, 1983], combines random and reservation access methods by providing N mini-slots in each slot for sending transmission requests by means of contention access and M mini-slots for sending data. The protocol assumes the presence of an "on-board" processing switch. The functional block diagram that depicts CRRMA is shown in *Figure 3-28*. The first portion of a slot in which a message is transmitted on the uplink channel is divided into N mini-slots, each of which is long enough to carry a transmission request packet. The second portion of the slot is intended for the transfer of one data packet. The duration of this portion of the slot is equivalent to M mini-slots (where M >> N). A station makes a request for transmission by sending a request packet in any of the N mini-slots chosen at random. The number of mini-slots, 'N', available in a slot is typically, an order of magnitude smaller than the number of stations that could possibly exist. Thus, there is a non-zero probability that requests for transmission reservations from multiple stations are made in the same mini-slot and thus, collide. An "on-board scheduler" (arbiter) on the satellite processes the valid requests. It sends control information specifying the status of a slot to indicate whether the slot has been reserved, through a downlink broadcast, to the earth stations. Every station that has a packet to transmit at the instance of a given slot (say the $i^{th}$ slot) determines a course of action in accordance with this control information. If the data portion of the slot is reserved for a station then the station holding the reservation transmits its packet. If a slot is not reserved (as indicated by the control information) the data portion of the slot is open for contention and multiple stations might attempt to transmit, in that slot. If the control information broadcast on the satellite downlink assigns the data portion of a slot to a station then the slot is said to be in a "reserved" state; otherwise it is in a "contention" state. If a request packet has collided then the retransmission of the request packet is necessary. Collisions (of either a request packet or a data packet) in a slot are detected after a round trip delay (the control information on the downlink facilitates this detection). Let the round trip delay be equal in duration to Q slots. If there was a collision of a request packet in the $i^{th}$ slot, then the mini-slots of the $(i + Q)^{th}$ slot are said to be in "retransmission" state. If there was no collision, then the slot is said to be in a "free" state. If the state of the $(i + Q)^{th}$ slot is "retransmission-contention", i.e., the mini-slots are in a "retransmission" state and data portion is in a "contention" state, then multiple (at least two) request packets have collided in the $i^{th}$ time slot. Thus, in this state, if the data portion is open for contention, then the transmitted packets are bound to collide again. Instead, in CRRMA, the M mini-slots of the data portion of the $(i + Q)^{th}$ slot are made available for request transmissions. Thus, the probability of collision amongst the competing stations is reduced.

When the on-board scheduler receives a successful request it stores this in a request queue that is serviced in accordance to a FCFS (first come first serve) policy. The access control information with regards to the slots that are reserved is relayed on the downlink in addition to any data that is to be transmitted, as shown in *Figure 3-28*. We reiterate that, the control information for determining the transmission in the $(i + Q)^{th}$ slot is triggered by transmissions in the $i^{th}$ slot.

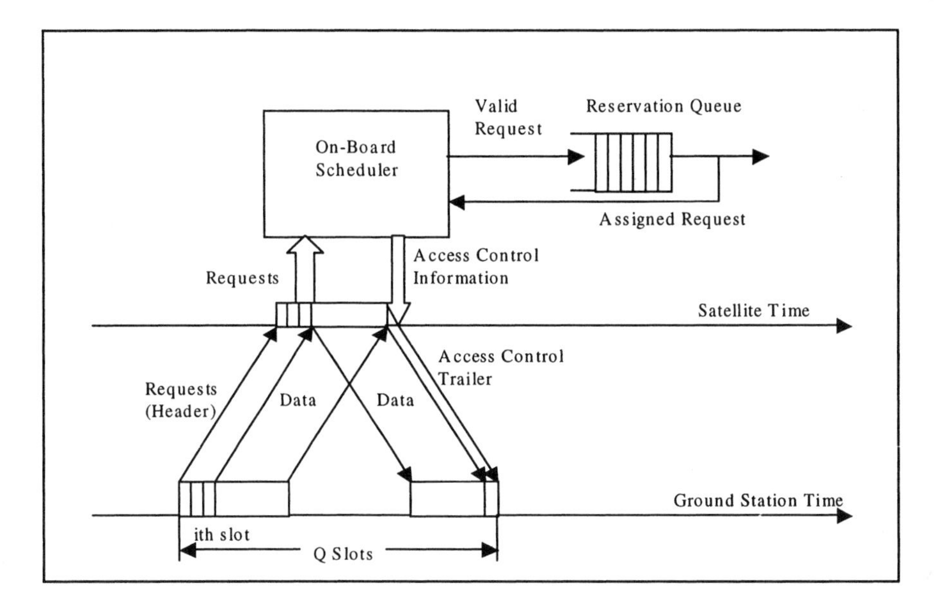


*Figure 3-28.* Functional block diagram for depicting CRRMA

CRRMA facilitates channel control by means of one of two algorithms. The channel can be controlled either by means of 'Uncontrolled Channel Access (UCA)' or 'Controlled Channel Access (CCA)'. In UCA stations transmit whenever they have a new packet for transmission and the slot is not reserved by any station. If a request packet in one of the mini-slots of the $i^{th}$ slot collides, then it is retransmitted in any of the mini-slots in the $(i + Q)^{th}$ slot. In CCA a request for the transmission of a newly arrived packet (for which no request has been made previously) is transmitted only in a "free" slot, i.e., it cannot be retransmitted in a slot that is in the "retransmission" state. This has the effect of reducing the probability of collisions in slots that are in the "retransmission" state. If there is a collision in more than one mini-slot, then the colliding stations are partitioned into groups and contention is resolved one group at a time. This resolution process is carried out by the on-board arbiter. UCA is simpler to implement but can become unstable if the number of mini-slots 'N' is small, as the stations transmit request packets for newly arriving packets even in "retransmission" slots. For lower values of 'N', the CCA algorithm performs better than the UCA algorithm. The CRRMA protocol provides a good delay-throughput performance for a wide range of loads.

# 7. CONCLUSIONS AND SUMMARY

Despite the fact that there is no protocol that performs better than the others for all types of traffic, given different combinations of offered traffic characteristics and required performance metrics, some protocols have characteristics that make them more suitable for satellite communications than others. In general, hybrid protocols that are designed to incorporate the advantages of both random and reservation based access have better throughput versus delay performance over a range of offered loads than either random or reservation access methods. They can also adapt more efficiently to network dynamics such as scalability and reconfigurability. The performance of these protocols is summarized in *Table 3-1*. This table is based on the assumption that the offered traffic is bursty and can dynamically change in time. The suitable MAC protocols for each possible offered traffic type are listed in *Table 3-2*.

*Table 3-1.* Performance comparison

| *Protocol* | *TDMA* | *G-TDMA* | *S-ALOHA* | *R-ALOHA* | *RRR* | *SRUC* |
|---|---|---|---|---|---|---|
| *Average Throughput* | Low | High | Low | High | High | High |
| *Mean Delay* | Low | Low | Very Low | Very Low | Very Low | Very Low |
| *Stability* | High | High | Low | Med | Med | High |
| *Scalability* | No | No | Yes | Yes | Limited | Yes |
| *Reconfigurability* | No | Low | High | High | High | High |
| *Broadband Applications* | Yes | Yes | No | No | Yes | Yes |
| *Cost/ Complexity* | Med | Med | Low | Low | Med | High |

For traffic characterized by short infrequent messages random access is the most appropriate. Random access protocols such as ALOHA, S-ALOHA, SREJ-ALOHA have the advantage of quick channel access at low loads. This accounts for the low delay that is achievable at low loads. Furthermore, this low delay is achievable even if the number of stations in the system is large. However, at higher offered loads there are an increased number of collisions, since now, there is a significant probability of multiple users trying to access the channel at the same time. Thus, the throughput goes down. This effect is shown in *Figure 3-29*.

If the stations generate primarily heavy loads of traffic or continuous traffic streams (or CBR type traffic), then a fixed allocation such as TDMA or FDMA may be appropriate. However, such static assignment of bandwidth may lead to under-utilization of the resources. Another option is to allocate bandwidth on-demand. In a demand assignment scheme, a station will have to send a message to the arbiter (which could be on-board the satellite or on the ground) to make a request for resources. Resources are thus, allocated on demand. Demand assignment is especially attractive when the stations are heavily loaded since, it might be expected that reservations will lead to finite delays. However, these schemes incur an initial delay of at least 270 milliseconds (for a geo-synchronous satellite) even at very light loads. Furthermore, the performance of the demand-assignment schemes heavily depend upon the methodology used for the arbitration of access to a separate channel (in time or frequency) that is used to transmit reservation requests. The performance of a typical demand-assigned multiple access scheme may be depicted as shown in *Figure 3-29.*

*Table 3-2.* Relation between traffic model and MAC choice

| *Traffic Model* | *MAC class choice* |
|---|---|
| Non-bursty traffic, CBR | Fixed assignment |
| Bursty traffic with short messages | Random access |
| Bursty traffic with long messages and with a large number of lightly loaded stations | Reservation protocols with contention based access for transmitting reservation requests |
| Bursty stations with long messages, and small number of possibly heavily loaded stations | Reservation protocols with fixed TDMA reservation channel |

There are schemes that incorporate the benefits of both random and reservation based access. These schemes attempt to retain the benefits of random access at low loads while emulating demand-assignment at high loads. Thus, they enjoy low delays at low loads and finite delays at high loads. They are suited for both short bursty traffic and for heavy sustained traffic. While the former of traffic is primarily transported by means of random-access, the latter is typically transported by using demand assignment. The expected performance of such a scheme (like CFDAMA) is depicted in *Figure 3-29.*

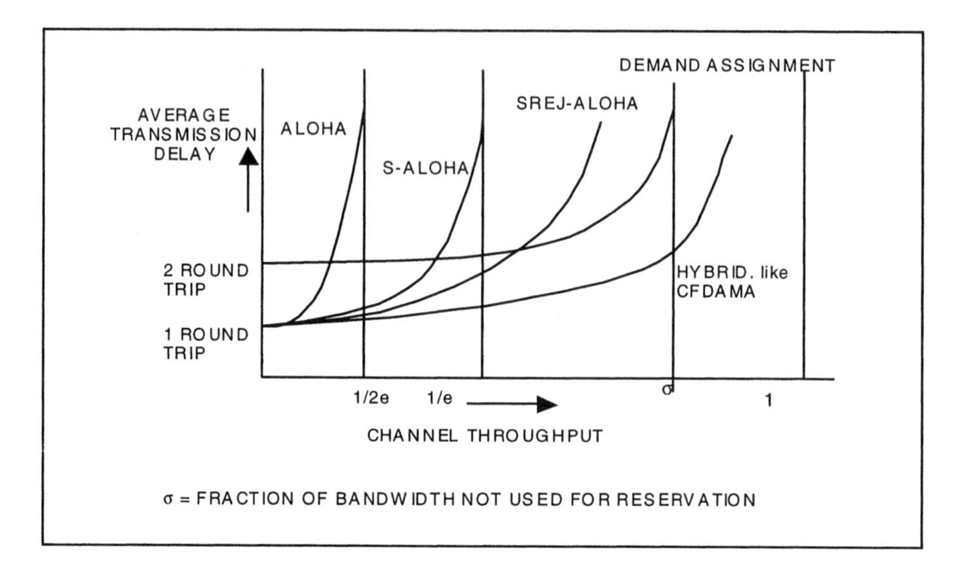


*Figure 3-29.* A Performance comparison of the various MAC protocol types

# REFERENCES


Bellini, S. and Borgonovo, P. (1980). On the throughput of an ALOHA channel with variable length packets. *IEEE transaction on Communications*, 28(11):1932-1935.

Berger, T. and Mehravari, N. (1981). An improved upper bound on the throughput of a multiaccess broadcast channel. In *Proceedings of IEEE Infocom. Theory Symp.*

Borgonovo, F. and Fratta, L. (1978). SRUC: A technique for packet transmission on multiple access channels. In *Proceedings of the ICCC*, pages.601–607.

Capetanakis, J. (1979). Tree algorithms for packet broadcast channels. *IEEE Transaction on Information Theory*,.IT-25(5):505-513.

Connors, D. and Pottie, G. (2000). Response initiated multiple access (RIMA), a medium access control protocol for satellite channels. In *Proceedings of Global Telecommunications Conference*, pages 1124-1129.

Jacobs, I., Binder, R., and Hoverstein, E. (1978). General purpose packet satellite networks. *Proceedings of IEEE*, 66(11):1448–1467.

Krishnamurthy, S. (1994). *Combined free/demand assignment multiple-access (CFDAMA) protocols for integrated data/voice satellite communications*. Master's Thesis, Department of Electrical and Computer Engineering, Concordia University, Canada.

Lam, S. (1980). Packet broadcast networks – A performance analysis of the R-ALOHA protocol. *IEEE Transactions on Computers*, 29(7):596–603.

Lee, H. and Mark, J. (1983). Combined random/reservation access for packet switched transmission over a satellite with on-board processing. I. Global beam satellite. *IEEE Transactions on Communications*, COM-31(10):1161-1171.

Le-Ngoc, T. and Krithnamurthy, S. (1996). Performance of combined free-demand assignment multiple-access schemes in satellite communications." *Int. J. Satellite Commun.*, 14(1):11-21.

Le-Ngoc, T. and Jahangir, I. (1998). Performance analysis of CFDAMA-PB protocol for packet satellite communications. *IEEE transactions on Communications*, 46(9):1206–1214.

Le-Ngoc, T. and Yao, Y. (1991). CREIR, a multiple access protocol for on-board processing satellite systems. In *Proceedings of Can. Conf. On ECE.*

Maral, G. and Bousquet, M., (1998). *Satellite Communications Systems (Third Edition)*, John Wiley & Sons.

Palmer, L. and White, L. (1990). Demand Assignment in the ACTS LBR System. *IEEE Transaction on Communication*, 38(5):684–692.

Pavey, C., Rice, Jr., R., and Cummins, E. (1986). A performance evaluation of the PDAMA satellite access protocol. In *Proceedings of IEEE INFOCOM,* pages 580-589.

Peyravi, H. (1995). *Multiple Access Control Protocols for the Mars Regional Network: A Survey and Assessments.* Technical report, Department of Math and Computer Science, Kent State University.

Peyravi, H. (1999). Medium Access Control Protocols Performance in Satellite Communications. *IEEE Communication Magazine*, 37(3):62–71.

Raychaudhuri, D. (1985). Announced retransmission random access protocols. *IEEE Transactions on Communications*, 33(11):1183–1190.

Roberts, L. (1975). *Dynamic allocation of satellite capacity through packet reservation.* Book chapter of *Computer Communication Networks*, edited by R.L. Grimsdate and F. F. Kuo, Noordhoff Internet Publishing.

Rom, R. and Sidi, M. (1990). *Multiple Access Protocols: performance and analysis,* Springer-Verlang, New York.

Stallings, W. (1997). *Data and Computer Communications (5th edition)*, Prentice Hall.

Suda, T., Miyahara, H., and Hasegawa, T. (1983). Performance evaluation of an integrated access scheme in a satellite communication channel. *IEEE Journal on Selected Areas in Communications*, SAC-1(1):153–164.

Tanenbaum, A. (1996). *Computer Network (Third edition)*, Prentice Hall.

Tobagi, F. (1980). Multiaccess protocols in packet communication systems. *IEEE Transactions on Communications*, COM-28(4):468–488.

Tobagi, F. and Kleinrock, L. (1976). Packet switching in radio channels: Part III — polling and (dynamic) split channel reservation multiple access. *IEEE Transactions on Communications*, 24(8):832–845.

Wieselthier, J. and Ephremides, A. (1980). A new class of protocols for multiple access in satellite networks. *IEEE Transactions on Automatic Control*, 25(5):865–879.

Yum, T. and Wong, E. (1989). The scheduled retransmission multi-access protocol for packet satellite communication. *IEEE Transactions on Information Theory*, 35(6):1319–1324.

Chapter 4

# DIRECT BROADCAST SATELLITES AND ASYMMETRIC ROUTING

Yongguang Zhang
*HRL Laboratories, LLC*

**Abstract** Advances in digital satellite communication technologies have allowed the use of broadcast-only satellites to provide Internet services. These systems, while having many unique benefits, bring about a new type of link technology – unidirectional links. Traditionally, most Internet routing protocol designs and their implementations assume that every link is bi-directional. With these broadcast-based satellite systems, the assumption is no longer true. In this chapter, we describe the challenges of dynamic asymmetric routing with unidirectional links, and explain a practical solution that has been adopted by the IETF (Internet Engineering Task Force).

**Keywords:** UDLR, Asymmetry, Routing, Direct Broadcast Satellites.

## 1. INTRODUCTION

Rapid advances in communication technologies in the 90s have allowed the use of one-way broadcast-only satellites (DBS or DVB-S) as a new networking medium for the Internet. These systems have a unique combination of compelling benefits: large geographical coverage, high bandwidth, an inherently broadcast nature, and low terminal costs. They are especially suitable for applications that require rapid deployable infrastructure and/or one-to-many communications [Padmanabhan et al., 1996; Zhang and Dao, 1996; Zhang et al., 1997]. Several commercial offerings have already been using such systems to provide Internet connectivity, such as DirecPC™ [1].

Nevertheless, these satellites have brought unconventional networking characteristics to the Internet. Traditionally, communication links in the Internet are duplex and bidirectional. Consequently, routing in the Internet is symmetric: the paths to and from a node are usually through the same set of routers. Now, these broadcast satellite links are inherently one-way only: communication can take place only from a small subset of nodes (called *feeds*) to rest of the nodes (called *receivers*). Feeds are equipped with large ground terminals that have adequate power to transmit data to the satellite for broadcast. Receivers only have small receive-only antenna; they can receive broadcast signal from the satellite but do not have the physical mean to transmit signals, they are hence not capable of reaching back to feeds. Such a communication network is unidirectional in nature. We call such a network link a *UniDirectional Link* (UDL). Besides satellites, certain cable modem systems are also UDLs.

UDL can be integrated into the Internet as a subnet technology, if there already exist other paths in the Internet from the receivers to the feeds [Falk, 1994; Arora et al., 1996]. Internet routers can be configured to use the UDL to deliver IP packets in one direction and use the alternative path for the other direction. As long as the routes are setup properly, the end-to-end TCP/IP applications will operate properly as they need not know how the IP packets are delivered and what paths are used. We called such routing scenario *asymmetric routing* or *UniDirectional Link Routing* (UDLR).

DirecPC™ is perhaps the first commercial UDL network. It relies on simple static configuration to set up the asymmetric routes. It is well understood that static routing has its limitations – routes need to be pre-configured, may not be optimal, and cannot adapt to topology changes. For example, in DirecPC network all traffic from DirecPC receivers has to take a detour to the feed before redirected to the destinations. To correct this, dynamic routing should be used. Unfortunately, dynamic UDLR is hard to achieve in today's Internet. We will explain this problem in the next section.

## 2. PROBLEMS WITH DYNAMIC ASYMMETRIC ROUTING

UDLR cannot be easily integrated with Internet's the dynamic routing infrastructure. Most dynamic routing protocol implementations make the assumption that links are bidirectional. If a node receives a routing information packet on one interface, the routing software often assumes the source (sender of the packet) is automatically reachable via that same interface. This assumption breaks down in a network with UDLs. In a network with UDLs, a packet containing the routing information might be received via one link but it might not be possible to transmit packets on that link.

### 2.1 Unicast

Many unicast routing protocols in the Internet use a distance-vector algorithm to compute dynamic routes [Huitema, 2000]. Before this algorithm can be applied to a network with UDLs, special care must be taken to propagate the routing information. Figure 4.1 illustrates this implication using a simple 3-

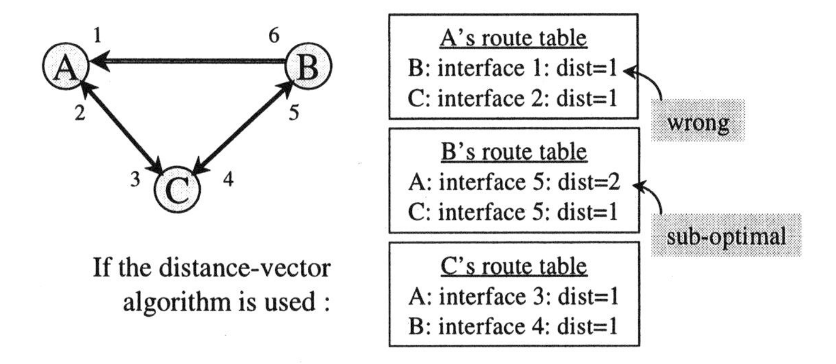


*Figure 4.1.* Applying distance-vector to UDL

node network with a UDL. Router A receives routing information from both B and C, however, router B only receives routing messages from node C. As shown in the figure, the route tables computed by the distance-vector algorithm are either incorrect or sub-optimal. The route table at A contains an incorrect entry for destination B as no packets can be sent out of interface 1 (the receive-only interface of the UDL receiver). Moreover, the route table at B contains a sub-optimal entry for destination A, via the regular bidirectional network which makes A two hops away, while the actual shortest path to A from B is via interface 6 (the send-only interface of the UDL feed).

Therefore the first problem involving unicast routing in a network with UDLs is that of sending packets out of the UDL send-only interface at the UDL sender. As that interface does not receive any routing packets, it will conclude that no networks can be reached via that interface. This is incorrect as all networks on the UDL receiver side of the network are in fact reachable via that interface and the link.

The second problem arises for packets sent from the UDL receiver to networks on the UDL sender side of the network. The UDL receiver receives routing information via the UDL. Therefore, it concludes that the shortest path back to the UDL sender, or any other network beyond the UDL sender is only one hop away and can be reached via the UDL receive-only interface. However, it is not possible to transmit packets from that interface as it is receive only. Therefore, all networks on the UDL sender side of the network become unreachable from all networks on the UDL receiver side of the network.

The correct route table in the context of this UDL should be one that causes packets sent from the UDL receiver to the UDL sender to go via the bi-directional network and packets sent from the UDL sender to the UDL receiver to go via the UDL.

## 2.2 Multicast

Several multicast routing protocols in the Internet use reverse path forwarding [Deering and Cheriton, 1990] to derive their multicast forwarding tables. Their reliance on unicast routing to aid multicast forwarding makes the multicast routing protocols vulnerable to the same problems as unicast routing protocols.

Figure 4.2 illustrates the problems with incorrect multicast routing in a network with UDLs. In the first scenario, the multicast source S is either at the UDL receiver, or on a network downstream to the UDL receiver. The shortest reverse path from the multicast receiver at the UDL sender or on a network upstream to the UDL feed, is via the UDL link. However, as the multicast data cannot be forwarded from the source S to the receiver C via the UDL, therefore, the multicast receiver will not be able to receive any multicast packets from the the multicast source S. Similarly, in the second scenario where the multicast source S is at the UDL sender or on a network upstream to the UDL sender, and the multicast receiver is downstream on the UDL receiver side, the shortest reverse path from the multicast receiver to the multicast sender will be via the bi-directional network, although the most efficient forwarding path for the multicast traffic from the source to the receiver is via the UDL. Therefore in both cases, using the shortest reverse path scheme results in incorrect multicast forwarding tables.

The correct multicast route table in the context of this UDL should forward multicast packets via the bidirectional network in the first scenario and via the UDL in the second scenario.

# 3. TUNNELING: A PRACTICAL SOLUTION

There are several possible directions to resolve this problem. For example, if we can modify all routing protocols to remove the protocol dependency on the bidirectionality of a link, we can safely integrate UDLs with the current Internet routing architecture. There were indeed a few proposals to change RIP and DVMRP for this purpose [Duros and Dabbous, 1996]. Another direction is to develop routing algorithms especially for UDLs. These algorithms can provide a general solution for routing in the Internet, because a bidirectional link is simply a set of two UDLs. However, these represent *long-term* solutions since they require significant changes to the existing Internet infrastructure (which we will have further discussion in Section 6).

To quickly integrate the emerging broadcast-based satellite networks into the existing Internet, a *short-term* solution was needed. It must accommodate UDLs into the current Internet infrastructure without changing it. One such approach is *tunneling* [Karir and Zhang, 1998]. If we assume that UDLs are simply overlays on top of a bidirectional base network, these UDLs can be made into virtual bidirectional links by each pairing with an IP tunnel from the

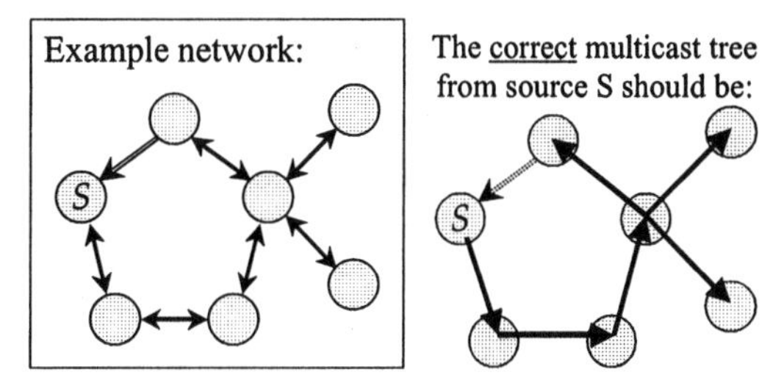


But, the actual tree built by reverse path forwarding:

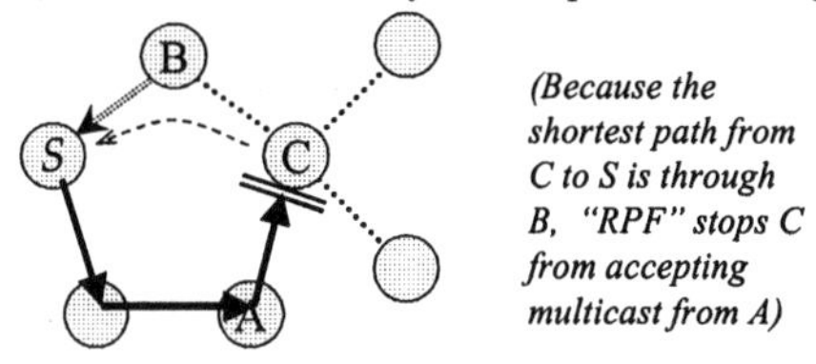


Scenario 1: source at downstream UDL

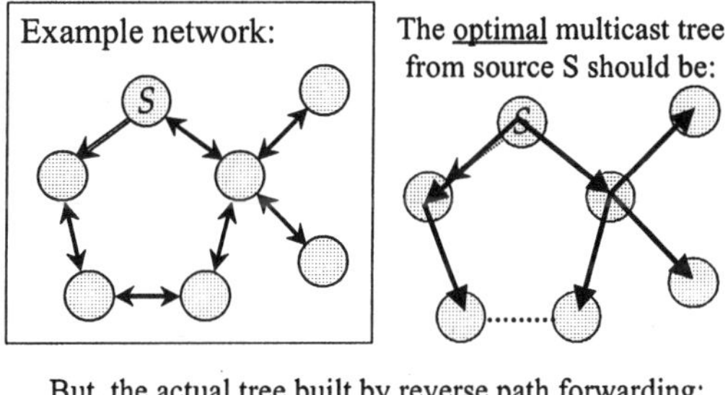


But, the actual tree built by reverse path forwarding:

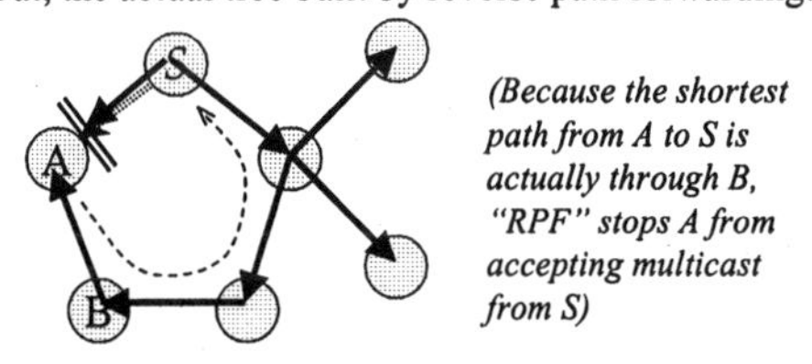


Scenario 2: source at upstream UDL

*Figure 4.2.* Applying reverse path forwarding to UDL

receiver to feed through the base network. This hides the asymmetry so that existing routing protocols can work.

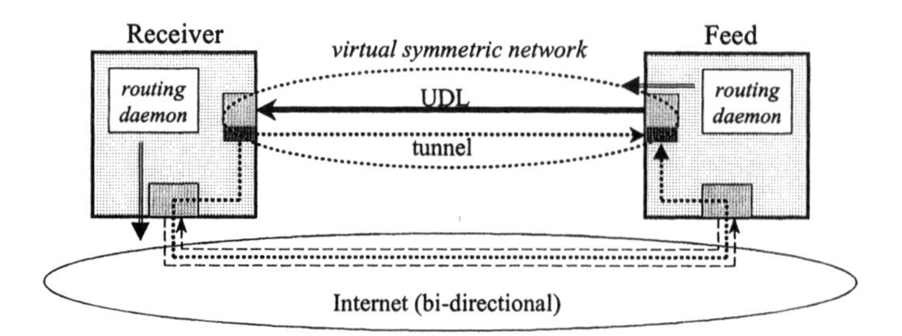


*Figure 4.3.* Approaching the UDLR problem with a tunneling mechanism

Figure 4.3 gives a high-level conceptual illustration. The IP tunnels create an illusion of a virtual bi-directional network for routing purposes. Each UDL receiver establishes a tunnel to the UDL feed via the existing bi-directional network. This tunnel carries encapsulated routing information directly from the UDL receiver back to the UDL feed. In this way the UDL sender and the UDL feed now have a virtual symmetric link between them through which they can exchange routing information. With carefully chosen routing metrics (see examples in next section), the data packets themselves do not flow through the tunnel, only the packets containing routing information go through the tunnel.

Experiments (such as [Karir and Zhang, 1998; Nishida et al., 1999]) have shown that this approach is easily deployable and is suitable as a practical solution to the UDLR problem. Although it makes the assumption that there must be a bidirectional base, this assumption matches the reality for broadcast-based satellite networks, many of which are indeed used as overlays and bypasses to enhance the current Internet performance.

## 4. DEMONSTRATION OF TUNNELING APPROACH

We now demonstrate the tunneling approach with a small experiment. The goal is to show how we can use tunnels to set up a network with UDLs with an existing routing protocol (RIP [Hedrick, 1988]).

Figure 4.4 describes the topology of the demonstration network. It consists of many Linux PCs acting as routers/hosts and several Ethernet segments to connect them. The network setup consists of a UDL subnet among three routers (A, B, C). Router C on the UDL has send-only capability and is the UDL feed, Router A and router B are on the receive-only side of the UDL and hence are the UDL receivers. Router A's receive-only interface is denoted as A.u (`192.168.4.1`), router B's receive-only interface is B.u (`192.168.4.4`), and router C's send-only interface is C.u (`192.168.4.2`). Since the UDL subnet is implemented by a bidirectional Ethernet segment, we emulate the UDL feature by running firewall utility on each UDL receiver (A and B). The firewalls are

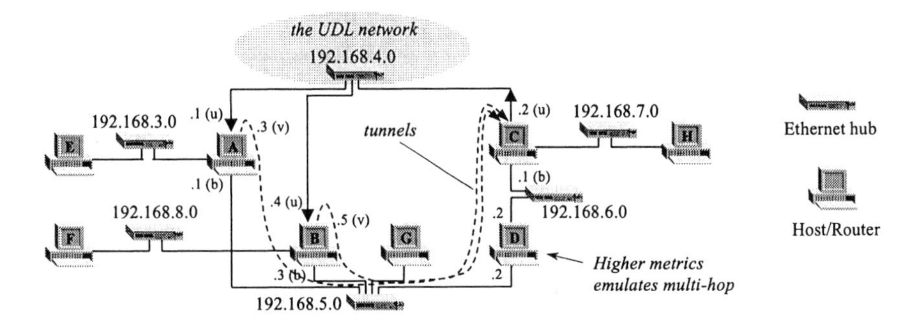


*Figure 4.4.* A demonstration network configuration

configured to prevent any packets from being sent out of the the UDL receive-only interfaces (A.u and B.u), which in effect provides us with a UDL where packets can only be sent from a UDL feed to a UDL receiver. In addition, router A, B, C all have other bidirectional interfaces, denoted as A.b (`192.168.5.1`), B.b (`192.168.5.3`), and C.b (`192.168.6.1`) respectively.

Router D acts as an emulated multihop bidirectional network. Router/host E, F, G, H are nodes that communicate with one other over this internetwork. In addition to the base topology shown in the figure, we can add and remove hosts and routers as required to demonstrate the various network topology. We can also supply artificial routes in the edge routers (E, F, G, H) to observe their propagation in a network with UDLs.

The first step is to set up a tunnel from a UDL receiver to a UDL feed. The goal is to make the tunnel as transparent as possible, so that we make little change to the routing software operation. Fortunately, many Unix operating systems support IP tunneling with a virtual network device, which delivers a packet by encapsulating it in another IP packet with a configurable destination address in the outer packet. In Linux, we can use the `tunl0` device.

Here, we need to create a tunnel interface from router A to router C, and another one from router B to router C. For example, the following Linux commands in router A establish the tunnel:

```
A>    ifconfig tunl0 inet A.v -pointopoint allmulti up
A>    route add -host C.u gw C.b dev tunl0
```

where A.v denotes the IP address assigned to the tunnel device (`192.168.4.3`). The first line sets up the network interface, the second line instructs that the route to the UDL feed (C.u) is via the tunnel (`tunl0`) to the UDL feed's bidirectional interface (C.b). Both lines together create a one-way tunnel from the UDL receiver to the feed.

The IP address assigned to the tunnel device can be from the same subnet as the UDL receiver interface, so that logically they can be treated as the same by the UDL feed. For example, when the routing daemon (at router A) sends out a message to the UDL feed (router C), it will carry the tunnel device IP address (A.v) as the source address. If the feed's routing software uses this address (A.v) to reply, the reply messages will go out the UDL send-only interface (C.u) because both are from the same subnet. However, the reply address (A.v) has no corresponding link-layer address because the virtual tunnel interface is not a physical device (unlike the receive-only interface A.u). To amend this, we set up an artificial entry at the ARP [2] table of the UDL feed so that the tunnel IP address (A.v) has the same link-layer address as its physical counterpart (A.u). Any messages from the UDL feed to the UDL receiver using the tunnel IP address (A.v) will be received by the physical receive-only interface (A.u). This can be easily accomplished with the following commands at router C:

```
C>    arp -s A.v ‘arp A.u‘
C>    arp -s B.v ‘arp B.u‘
```

With these IP tunnels, we can now experiment with routing. Here the routing protocol is RIP and the routing software is `gated`. The protocol works as follows. `gated` periodically sends routing update messages on all interfaces. The routing update is in form of "distance-vector" – the number of hops to reach each stub network in the Internet. When `gated` receives a routing update message from another `gated`, it adds the cost metric of the receiving interface (usually 1) to the distance-vector and merges it into the its unicast route table. `gated` then updates the kernel route table.

In order to create the illusion of a virtual bi-directional network around the UDL, we have to use the tunnel to carry RIP routing updates from the UDL receiver to the UDL feed so that they arrive with unmodified route metrics. With a tunnel set up as described above, `gated` can recognize the tunnel interface and send out RIP protocol messages on the interface. The following line is added to `gated.conf` at router A so that the RIP messages are addressed from A.v to C.u:

```
sourcegateways C.u
```

These messages will be automatically encapsulated by the tunnel interface and appear at the UDL feed with unmodified metrics as desired.

Upon receiving the RIP messages, the UDL feed (router C) will recognize that the shortest path to hosts downstream the UDL receiver side of the network is via the RIP message sender (i.e. A.v). The UDL feed (Router C) will then start forwarding IP packets downstream UDL, using A.v as the next hop. Using the ARP table entry as described previously, we map these packets to the correct actual interface (A.u) that these packets must arrive. With the setup above the route table obtained at the UDL feed will be correct in the UDL context, and packet from the UDL feed to the UDL receiver will be forwarded via the UDL.

Normally, when a UDL receiver receives RIP routing update messages with a small metric via the UDL interface, it concludes that the UDL receive-only interface would be used to reach back to the UDL feed and some upstream networks. To prevent this wrongful setting, `gated` at the UDL feed must use a large metric when it send RIP update messages to the UDL subnet. This metric should be specified as being larger than the sum along the bidirectional path from the UDL feed to the UDL receiver, so that the UDL receiver can logically conclude that the best path to the UDL feed or upstream networks is through the bidirectional network interface. This can be implemented easily by changing the metric value in UDL feed's `gated` configuration file, for example, the following line at router C's `gated.conf`:

```
interface C.u ... metricout 15
```

In practice, one can also instruct UDL feed's `gated` not to send out RIP information for the UDL subnet.

With this minor configuration setup, `gated` was able to generate correct route tables at all routers, and we were able to receive packets at the UDL feed, from the UDL receiver, via the bi-directional network.

## 5. RFC 3077: THE IETF STANDARD

By 1996, several research groups around the world had recognized the UDLR problem and started working on short-term solutions that did not require significant changes to the existing Internet hardware/software infrastructure. They included INRIA (France), WIDE (Japan), Hughes Research Labs (now known as HRL Labs, USA), TRW (USA), and others. To avoid duplicated work and incompatible solutions, researchers came together at the IETF (Internet Engineering Task Force, an Internet standardization body) to formalize this problem and to standardize a solution. An IETF working group called UDLR was created in 1997 as the result of this collaboration. Since there were already several studies and they all pointed to tunnelling as the right direction, the discussions in the working group were mainly focused on the details of the architecture and protocols. Finally, the working group produced an IETF standard document – RFC 3077 [Duros et al., 2001] – as the recommended solution to the UDLR problem in the Internet. In March 2001, the IETF has approved this document as a proposed Internet standard.

The IETF standard solution is very similar to the tunneling approach described in the previous section. Here, the receiver uses a link-layer tunneling mechanism to forward packets to feeds over a separate bidirectional IP network (i.e., the base). Titled "A Link-Layer Tunneling Mechanism for Unidirectional Links," RFC 3077 describes in detail the mechanism to emulate full bidirectional connectivity between all nodes that are directly connected by a UDL.

## 5.1 Topology and Requirements

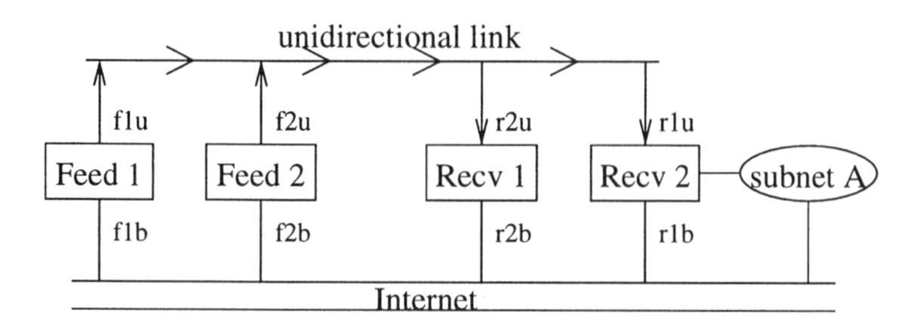


*Figure 4.5.* Generic topology for RFC 3077

Figure 4.5 depicts a generic topology for RFC 3077. It is very similar to what we have described in the previous section (c.f. Figure 4.3). It uses the same terminology such as "UDL", "feed", and "receiver", and it further divides "feed" into two categories: *send-only feed* that has send-only connectivity to a UDL and *receive-capable feed* that has both send and receive connectivity to the UDL. That is, some feeds may have the receiver function themselves, enabling them to receive from the same UDL.

As depicted in Figure 4.5, in addition to the UDL interface (f1u, f2u, r1u, r2u), both feeds and receivers have a second interface (f1b, f2b, r1b, r2b). This second interface connects them to a well-connected bidirectional infrastructure network. Tunnels will be constructed through this bidirectional network. This infrastructure network is assumed to be the Internet. This second interface is called the "bidirectional" interface.

Here, f1u, f2u, r1u, and r2u represent the IP address of the UDL interfaces for Feed 1, Feed 2, Receiver 1, and Receiver 2, respectively. And f1b, f2b, r1b, and r2b represent the IP address of their respective bidirectional interfaces to the Internet. Subnet A denotes a local area network connected to Receiver 1.

The tunneling mechanism will emulate a broadcast bidirectional network from a UDL. The reason that it specifies a broadcast capable link-layer is to allow the immediate deployment of existing higher level protocols without change. Though other network structures, such as NBMA (non-broadcast multiple access [Przygienda et al., 2000]), could also be emulated, a broadcast network is more generally useful. For example, a layer-3 network could be emulated but a link-layer network allows the immediate use of any other network layer protocols, and most particularly allows the immediate use of ARP.

To define what functionalities this tunneling mechanism has to perform, RFC 3077 classifies the communications in a broadcast network into following scenarios:

- Scenario 1: A receiver can send a packet to a feed (point-to-point communication between a receiver and a feed).
- Scenario 2: A receiver can send a broadcast/multicast packet on the link to all nodes (point-to-multipoint).
- Scenario 3: A receiver can send a packet to another receiver (point-to-point communication between two receivers).
- Scenario 4: A feed can send a packet to a send-only feed (point-to- point communication between two feeds).
- Scenario 5: A feed can send a broadcast/multicast packet on the link to all nodes (point-to-multipoint).
- Scenario 6: A feed can send a packet to a receiver or a receive capable feed (point-to-point).

These scenarios are all possible on a broadcast network. Scenario 6 is already feasible on a UDL. The link-layer tunneling mechanism should therefore provide the functionality to support Scenarios 1 to 5.

## 5.2 Tunneling Mechanism Details

The link-layer tunneling mechanism operates underneath the network layer and becomes transparent to it: the link appears and behaves to the network layer as if it was bidirectional. For example, Figure 4.6 depicts a layered representation of how the tunneling mechanism operates to support the above Scenario 1.

On the receiver side, when a higher layer packet (e.g., IP or ARP) is delivered to the link layer of the UDL interface for transmission, it will be first encapsulated in a LL (link-layer) frame [3] whose header corresponds to the UDL interface. Since this frame cannot be sent directly over the UDL link, it is further encapsulated within an IP packet whose destination IP address is the feed's FBIP ("Feed Bidirectional interface IP address", like f1b or f2b in Figure 4.6). This IP address is called the tunnel end-point. The choice of tunnel end-point depends on the encapsulated frame's destination LL address. If it is the LL address of a feed interface connected to the UDL (Scenario 1), the tunnel end-point will be the feed's FBIP (f1b). Otherwise (either a LL broadcast/multicast address as in Scenario 2 or a non-feed receiver's LL address as in Scenario 3), the tunnel end-point will be the "default" feed's FBIP.

On the feed side, when a packet is delivered to the UDL link-layer interface for transmission, its treatment also depends on the destination's corresponding LL address. If it is the LL address of a receiver or a receive-capable feed (Scenario 6), the packet is sent over the UDL in a classical IP forwarding process.

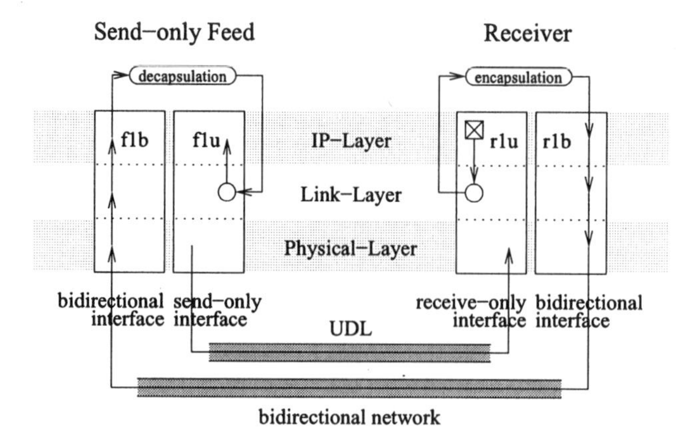


*Figure 4.6.* Scenario 1 using the link-layer tunneling mechanism

If it is the LL address of a send-only feed (Scenario 4), the packet is encapsulated and sent to the send-only feed's FBIP. Or, if it is a broadcast/multicast destination (Scenario 5), the packet is also sent over the UDL, but concurrently a copy of this packet is encapsulated and sent to every feed's FBIP in the list of send-only feeds. This way, the broadcast/multicast packet will reach all receivers and all send-only feeds on the UDL.

When an IP packet with encapsulated payload reaches its tunnel end-point at a feed, it arrives at the bidirectional interface and traverses its way up the IP stack. When it enters the decapsulation process (see Figure 4.6), the original LL frame will be recovered intact. Further actions depend on the recovered frame's destination LL address. If it is the LL address of the feed's UDL interface (Scenarios 1 and 4), the frame will be passed to the UDL interface's link layer and traversing up to higher layers, creating the effect as if it were coming from the UDL and being delivered locally. If the destination LL address is a receiver's address (Scenario 3), the frame will be passed to the UDL interface's link layer and sent over to the UDL to the intended receiver – a receiver can now send packets to another receiver this way. Finally, it is slightly complicated for broadcast or multicast frames. If this feed receiving this frame is the designated multicast router (in charge of forwarding multicast/broadcast frames to all nodes), it tunnels this frame again to the list of send-only feed, forwards it directly over the UDL to all receivers and receive-capable feeds, and delivered it locally. Otherwise, the frame is only delivered locally.

## 5.3 Dynamic Tunnel Configuration

Each receiver needs to properly configure and maintain its tunnels to the feeds. First, it needs to know the tunnel end-point IP address in order to create and forward encapsulated packets (Scenario 1 and 4). It also needs to detect events like UDL link status changes (such as links going up or down) and feed status changes (such as feeds going up or down). In a typical UDL network like the satellite network, there are usually a small number of feeds (in many cases, a single feed) and a large number of receivers. For obvious reasons, we should not relay on manual configuration. The receivers should configure and maintain tunnels dynamically, both for scalability, and in order to cope with the following events:

1 New feed detection. When a new feed comes up, every receiver must create a tunnel to enable bidirectional communication with it.

2 Loss of UDL detection. When the UDL is down, receivers must disable their tunnels. The tunneling mechanism emulates bidirectional connectivity between nodes. Therefore, if the UDL is down, a feed should not receive packets from the receivers. Protocols that consider a link as operational if they receive packets from it (e.g., the RIP protocol) require this behavior for correct operation.

3 Loss of feed detection. When a feed is down, receivers must disable the corresponding tunnel. This prevents unnecessary packets from being tunneled to an unreachable destination. For instance, there is no need for receivers to forward a broadcast message through a tunnel whose end-point is down.

To assist the tunnel setup and maintenance task, RFC 3077 specifies a protocol called DTCP (Dynamic Tunnel Configuration Protocol). It provides a mean for receivers to dynamically discover the presence of feeds and to maintain a list of operational tunnel end-points. With DTCP, feeds periodically announce their tunnel endpoint addresses over the UDL. Receivers listen to these announcements and maintain a list of tunnel end-points.

The announcement is in form of a HELLO message. Figure 4.7 specifies the message format and information layout. It contains the following important information useful in tunnel setup:

- Command field (Com) – whether the feed is up and running (JOIN) or is being shut down (LEAVE).

- Nb-of-FBIP – number of bidirectional interface(s) that the feed has and that can be used as a tunnel end-point.

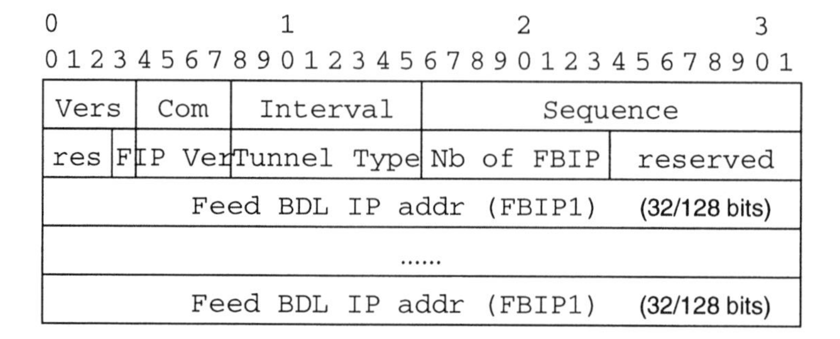


*Figure 4.7.* DTCP HELLO message packet format

- FBIP1, ..., FBIPn – A list of IP address(es) for the feed's bidirectional interface(s).
- Sequence – incremented whenever the feed changes the FBIP list.
- Tunnel type (see next subsection).
- F-bit – whether the feed is send-only or receive-capable.

The DTCP protocol runs on top of UDP. It uses multicast address `224.0.0.36` on UDP port number `652`. These two numbers have been assigned by IANA[4] for DTCP use only. Packets are sent to the `224.0.0.36` on port `652` over the UDL with a TTL (time-to-live) of 1. The packet source address is set to the IP address of the feed's interface connected to the UDL, which is referred as FUIP ("Feed Unidirectional IP address").

As long as a feed is up and running, it periodically announces its presence to receivers by sending a HELLO packet containing a JOIN command. Taking the example in Figure 4.5, Feed 1 sends HELLO messages with the FBIP1 field set to f1b and the source address FUIP being f1u. When a feed is about to be shut down, it should stop sending any more HELLO messages except a last one containing a LEAVE command, to inform receivers that this feed is no longer serving this UDL.

When a receiver is started, it joins the same multicast group (`224.0.0.36`) at the UDL and listens to incoming packets on the same port (`652`). Based on the reception of HELLO messages, receivers discover the presence of feeds, maintain a list of active feeds, and keep track of the tunnel end-points for those feeds. For each active feed, receivers keep an entry of the following information: FUIP, FUMAC (LL address corresponding to FUIP), FBIP1 to FBIPn, tunnel type, Sequence number (of the last HELLO received from this feed), and a timer used to time out this entry.

Upon the reception of a HELLO message with JOIN command, a receiver first verifies if the address FUIP already belongs to the list of active feeds. If it does not, the receiver will create a new entry from the information in the HELLO message and add to the list of active feeds. Otherwise, the receiver next compares the sequence number from the previous value from the same feed. If they are equal, it simply resets the timer. Otherwise all the information corresponding to FUIP is set to the new values from the HELLO message.

Upon receiving a HELLO message with LEAVE command from a feed, or when timeout occurs on the corresponding entry (meaning either the feed went down abruptly or the UDL is down), the bidirectional connectivity can no longer be ensured between the receiver and the feed. The receiver thus remove the corresponding entry from its list of active feeds. As a result, the list only contains operational tunnel end-points.

Hence, the DTCP protocol provides receivers with a list of operational feeds, and a list of usable tunnel end-points (FBIP1,..., FBIPn) for each feed. Taking the example in Figure 4.5, both receivers (Receiver 1 and Receiver 2) have a list of active feeds containing two entries: Feed 1 with FUIP being f1u and a list of tunnel end-points (f1b); and Feed 2 with FUIP being f2u and a list of tunnel end-points (f2b).

With the above tunnel configuration information, each receive needs to pick a feed and a tunnel end-point to set up the tunnel. In the most common case where each UDL has only one feed and one FBIP, the choice is simple. However, in more complex cases, the choice of the default feed need to be made independently at each receiver and this is often a matter of local policy, such as choosing the one with the lowest round trip time. Choosing a tunnel end-point from within the list (FBIP1 to FBIPn) is also a matter of local decision, although the "preferred" one is generally FBIP1.

## 5.4 Tunneling Protocol

The tunneling mechanism operates at the link layer and emulates bidirectional connectivity amongst receivers and feeds. For the forward direction along the UDL, a feed can usually send a frame to a particular receiver using a unicast LL destination address or to a set of receivers using a broadcast/multicast destination address. We should make sure that a similar way exist on the reverse direction. That is, a receiver should be capable of sending unicast and broadcast LL frames via its tunnels, and the feed should be able to process the incoming frame as if it came directly from the UDL (See Figure 4.6).

For implementation of receiver-to-feed tunneling, there are several tunneling protocols from which one can choose, such as IP-in-IP and GRE (Generic Routing Encapsulation [Farinacci et al., 2000]). The feed's local administrator decides what encapsulation it will demand that receivers use, and sets the tunnel type field in the DTCP HELLO message appropriately. RFC 3077 recommends

GRE because it supports the encapsulation of arbitrary packets/frames and allows the use of IP as the delivery protocol.

A GRE packet is composed of a header in which a type field specifies the encapsulated protocol (ARP, IP, IPX, etc.) [Farinacci et al., 2000]. In our case, only support for the LL addressing scheme of the UDL must be implemented. A packet/frame tunneled with a GRE encapsulation has the following format: the delivery header is an IP header whose destination is the tunnel end-point (FBIP), followed by a GRE header specifying the link layer type of the UDL. Figure 4.8 presents the format for the entire encapsulation packet (assuming no optional GRE header field).

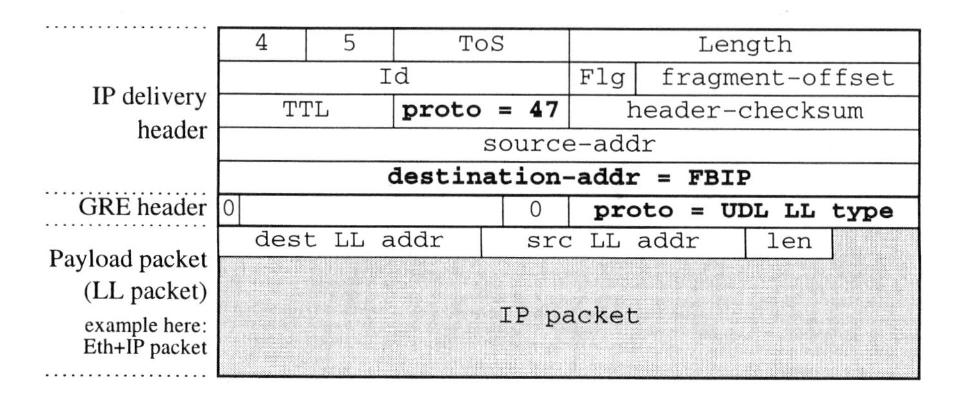


*Figure 4.8.* UDL encapsulation packet format

## 5.5 Current Status

Since its standardization process in IETF, there have been several implementations based on the tunneling mechanism described in RFC 3077 (such as [Izumiyama et al., 2001]). Many vendors have already included this feature in their commercial satellite and/or Internet products, such as Hitachi[5], UDcast[6], IPricot[7], and Thomcast[8]. These implementations are fully interoperable and are currently being used in several operational environments. The tunneling mechanism has then been recognized as a simple and yet standard solution to turn broadcast-based unidirectional satellite communications into bidirectional networks. It supports any existing return channel technology and support all Internet protocols.

After publication of RFC 3077 in 2001, the UDLR working group at IETF has since re-chartered itself to produce new RFCs that describe in detail the requirements of example case study scenarios such as configuration of multicast/unicast routing protocols over a UDL. The working group's web site is currently hosted at `http://www.udcast.com/udlr` .

# 6. LIMITATIONS AND LONG-TERM SOLUTIONS

The tunneling mechanism hides the unidirectional characteristics of the Internet link from the upper layers, so software and applications that operate in those layers can continue to operate under the presumption that the network underneath is bidirectional. However, not all link characteristics can be hidden easily, such as the delay of a satellite link or the bandwidth asymmetry. As a result, upper layer software unaware of this special link may still behave in an unexpected way. For example, a proper routing protocol should advertise a low cost metric along the unidirectional satellite link to encourage its use, but advertise an infinite metric along the reverse direction to discourage normal (non-routing-related) data traffic from going back through the tunnel. If such configuration is not properly set up, or if the route protocol use the same metric for both directions, the traffic may not go as planned.

For another example, some routing protocols require hop-by-hop acknowledgements, i.e., every routing message is to be acknowledged by the direct neighbor. Such acknowledgements for routing messages from a UDL feed will have to travel a long path back through the tunnel, leading to overhead.

Furthermore, the tunneling approach requires a well-connected bidirectional base network so that tunnels can be established easily. If such a base network does not exist, or the base network itself has UDLs, we may have a bootstrapping problem – both sets of UDLs may need each other to set up proper routing. While this is not a likely scenario with today's applications of satellite networks over the Internet, it does make an interesting long-term problem to consider asymmetric routing protocols that assume every link is a UDL. Since the basic techniques used by most current routing protocols are relatively simple and all have limitations with UDL networks, we may need to return to graph theory for more elaborated algorithms [Ernst and Dabbous, 1997; Bao and Garcia-Luna-Aceves, 1999].

Incidentally, researchers in wireless ad-hoc networks (where an all-wireless multihop network is formed by mobile computer nodes communicating through radios and relaying for each others who are out of radio range) are also working on the asymmetric routing problem. In wireless communications, nodes transmitting at higher power can have larger transmission ranges. A UDL will exist if nodes have different transmission ranges and one node is within range of the other but not vice versa. Although routing in wireless ad-hoc networks is very different from routing in the Internet, algorithms developed there might inspire long-term solutions to the UDLR problem in satellite networks.[9]

## Notes

1. DirecPC™ (`http://www.direcpc.com/`) is a satellite-based Internet service from Hughes Network Systems (`http://www.hns.com/`).

2. ARP (Address Resolution Protocol) is a commonly used mechanism to establish mapping between IP-layer and link-layer addresses [Huitema, 2000].

3. We use the term "LL" (link-layer) instead of "MAC" (media access control) to specify the ISO layer 2 framing and addressing. These two terms are often interchangable (especially for Ethernet) but are not always the same. We also use the word "frame" to refer link layer packets.

4. IANA: Internet Assigned Numbers Authority – a central coordinator for the assignment of unique parameter values for Internet protocols.

5. Hitachi claims its satellite router With-IT SAT1 is "the world's first-ever router" to employ RFC 3077: `http://www.hitachi-it.co.jp/english/withit/sat1.htm`.

6. A spin-off from INRIA for UDLR business development: `http://www.udcast.com/`

7. Satellite IP Router-Receiver IPR-Sc: `http://www.ipricot.com/broadbandprod/sc.htm`

8. Thomcast OpenMUX Galaxy product family: `http://www.thomcast.com`

9. Asymmetric routing in wireless ad-hoc networks is an active research topic but is outside the scope of this book. It is impossible to list all the references, but readers who are interested can get some ideas in these two papers [Bao and Garcia-Luna-Aceves, 1999; Prakash, 2001].

# References

Arora, V., Suphasindhu, N., Baras, J. S., and Dillon, D. (1996). Effective extensions of internet in hybrid satellite-terrestrial networks. In *Proceedings of the 1st Conference of Commercial Development of Space, Part One*, pages 339–344.

Bao, L. and Garcia-Luna-Aceves, J. J. (1999). Link-state routing in networks with unidirectional links. In *Proceedings of the Eight International Conference on Computer Communications and Networks*, pages 358–363.

Deering, S. and Cheriton, D. (1990). Multicast routing in datagram internetworks and extended LANs. *ACM Transactions on Computer Systems*, 8(2):85–110.

Duros, E. and Dabbous, W. (1996). Supporting unidirectional links in the Internet. In *Proceedings of the First International Workshop on Satellite-based Information Services*.

Duros, E., Dabbous, W., Izumiyama, H., Fujii, N., and Zhang, Y. (2001). A link-layer tunneling mechanism for unidirectional links. RFC 3077, the Internet Engineering Task Force.

Ernst, T. and Dabbous, W. (1997). A circuit-based approach for routing in unidirectional links networks. Technical Report RR-3292, INRIA Sophia Antipolis (URL: http://www-sop.inria.fr/rapports/sophia/RR-3292.html).

Falk, A. D. (1994). System design for a hybrid network data communications terminal using asymmetric TCP/IP to support internet applications. M.S. Thesis (CSHCN MS 94-2), University of Maryland.

Farinacci, D., Li, T., Hanks, S., Meyer, D., and Traina, P. (2000). Generic Routing Encapsulation (GRE). RFC 2784, the Internet Engineering Task Force.

Hedrick, C. L. (1988). Routing information protocol. RFC 1058, the Internet Engineering Task Force.

Huitema, C. (2000). *Routing in the Internet*. Prentice Hall, 2nd edition.

Izumiyama, H., Takei, J., Fujieda, S., Nishida, M., and Murai, J. (2001). A link-layer tunneling mechanism for unidirectional links. *IEICE Trans. Communications*, E84-B(8):2058–2065.

Karir, M. and Zhang, Y. (1998). An experimental study of asymmetric routing. In *Proceedings of the Third International Workshop on Satellite-based Information Services.*

Nishida, M., Kusumoto, H., and Murai, J. (1999). Dynamic tunnel configuration for network with uni-directional link. In *Proceedings of the 14th International Conference on Computer Communication.*

Padmanabhan, V., Balakrishnan, H., Sklower, K., Amir, E., and Katz, R. (1996). Networking using direct broadcast satellite. In *Proceedings of the First International Workshop on Satellite-based Information Services.*

Prakash, R. (2001). A routing algorithm for wireless ad hoc networks with unidirectional links. *ACM/Baltzer Wireless Networks Journal*, 7(6):617–626.

Przygienda, T., Droz, P., and Haas, R. (2000). OSPF over ATM and Proxy-PAR. RFC 2844, the Internet Engineering Task Force.

Zhang, Y. and Dao, S. (1996). Integrating direct broadcast satellite with wireless local access. In *Proceedings of the First International Workshop on Satellite-based Information Services*, pages 24–29.

Zhang, Y., DeLucia, D., Ryu, B., and Dao, S. (1997). Satellite communications in the global Internet: Issues, pitfalls, and potential. In *INET'97.*

# Chapter 5

# USING SATELLITE LINKS IN THE DELIVERY OF TERRESTRIAL MULTICAST TRAFFIC

Kevin C. Almeroth
*Department of Computer Science, University of California at Santa Barbara*

**Abstract** Multicast has been deployed and operating in the Internet since 1992. Through the provisioning of a multicast delivery service, researchers have been learning how to deliver scalable, one-to-many content. While much of the focus on multicast research continues to be on addressing the deployment challenges for the traditional Internet, other "non-traditional" infrastructures, like satellite, cable, and wireless networks are beginning to be recognized as important environments in which to provide and use multicast communication. The inherent broadcast nature of these networks suggests that multicast might be easier to provide as compared to the traditional Internet. In this chapter, we first examine how to configure a satellite network to support IP multicast. Next, we examine how to bridge Internet-based multicast sessions to a satellite network. Then, we report on our efforts to demonstrate the basic feasibility of delivering multicast sessions over a satellite network. Further, we focus on a quantitative analysis of the impact of satellite links on three network performance metrics: loss, jitter, and delay. In the final analysis, we answer the question on how and when to use satellite links in combination with terrestrial multicast-based sessions.

**Keywords:** multicast, routing, satellite, gateway, end-to-end performance

## 1. INTRODUCTION

A multicast service has been operating in the Internet since 1992. The first experimental use of multicast was over an overlay network called the Multicast Backbone (MBone). Through this experimental network, researchers have been learning how to deliver a scalable, one-to-many multicast service. The deployment of a truly scalable multicast service in the Internet has been one of the biggest challenges for researchers. The key reason can be found in the characteristics of the Internet itself. The bottom line is that the Internet provides a very limited delivery service; packets are carried best-effort across several administrative boundaries which are often congested and unstable, and can be subject to frequent routing pathologies. Packet loss rates, delays, and jitter can be high and conditions can vary for receivers in even small multicast groups.

While much of the focus on multicast research continues to be on addressing the challenges of the traditional Internet, other "non-traditional" infrastructures, like satellite, cable, and wireless networks are beginning to be recognized as important environments in which to investigate the provision of a multicast delivery service. The inherent broadcast nature of these non-traditional networks suggests that multicast might be easier to provide in these networks compared to the traditional Internet. As proof of this possibility, commercial vendors are finding they can offer some types of multicast services more efficiently over a customer's satellite networks. However, there are still a number of issues, especially relating to cost and service delivery, that remain unanswered.

There are a variety of applications that can be provided using multicast, including one-to-many software distribution, cache updates, database replication, streaming multimedia, multi-user games, interactive conferencing, etc. These classes of applications have different protocol requirements including real-time delivery constraints, jitter tolerances, and reliable data transfer requirements. Our focus is on the delivery of multicast-based applications over satellite links. The particular types of sessions we believe can be most efficiently delivered include those that: (1) cover a large geographic area; (2) include more than just a few receivers; (3) are of a broadcast nature with only a single or a few well-known transmitters; and (4) use streaming data types like audio/video/whiteboard/text. The prototypical example that we use is the streaming delivery of audio and video from the Internet Engineering Task Force (IETF)[1] meetings which occur three times a year at locations around the world.

In this chapter, we also report on our efforts to demonstrate the basic feasibility of delivering multicast sessions over a satellite network during 1997-1998. We were able to successfully take the audio and video from the 40th IETF meeting in Washington DC and send it over a satellite network. The next step is to answer the question of whether a satellite network can offer performance and cost advantages, and for what types of sessions would satellites offer the most significant advantages. To a certain extent, researchers already have a qualitative understanding of the issues. However, the focus here is to conduct a more quantitative analysis. The quantitative analysis is based on three metrics: loss, jitter, and delay.

This chapter is organized as follows. Section 2 gives an overview of multicast in the Internet. Section 3 describes how to deliver multicast data over satellite. Section 4 details our efforts to deliver multicast sessions over a satellite network. Section 5 describes our efforts to quantify the advantages and disadvantages of using a satellite network. Finally, section 6 lists the characteristics of multicast sessions most suitable for delivery via satellite links.

## 2. OVERVIEW OF MULTICAST DEPLOYMENT

Multicast is currently deployed over a fairly wide portion of the Internet. However, it still has yet to be deployed over a *large* portion of the Internet [Rajvaidya and Almeroth, 2001]. Still, multicast has evolved substantially from its first inception as an overlay network. While there is more thorough work describing the evolution of multicast[Almeroth, 2000], in this section we give only a brief overview.

Multicast deployment began as has been the Multicast Backbone (MBone). The MBone[Casner, 1994] is an experimental virtual network created to provide the means for multicasting data to any number of connected hosts. In 1992, 20 sites connected together via a rudimentary multicast network and received audio transmissions from the March meeting of the Internet Engineering Task Force (IETF)[Casner and Deering, 1992]. That first audio conference, carried over the Internet and the MBone, allowed a few members spread all over the world to hear what was being said in San Diego, California.

As multicast has evolved, new protocols have been deployed. In addition to the original Distance Vector Multicast Routing Protocol (DVMRP)[Deering and Cheriton, 1990], Protocol Independent Multicasting (PIM)[Deering et al., 1996] and Core Based Trees (CBT)[Ballardie et al., 1995] have been proposed. Currently the multicast infrastructure uses the Multicast Border Gateway Protocol (MBGP)[Bates et al., 1998], the Protocol Independent Multicast (PIM) protocol[Deering et al., 1996] and the Multicast Source Discovery Protocol (MSDP)[Farinacci et al., 1998]. MBGP acts as a route exchange protocol and allows propagation of topology information between domains. PIM uses route information to create, manage and propagate information about distribution trees. These trees consist of a set of forwarding table entries used for distribution of data from a particular source to the corresponding groups. Finally, MSDP acts as a source announcement protocol and is responsible for propagating information about active sources across the entire infrastructure.

As the MBone has evolved it has seen an increasingly diverse set of applications. Since the first audio conference in 1992, multicast has seen the development of new applications using audio, video, whiteboard, and text as media. Even more recently developed applications using non-streaming media and require additional network services like multicast-based congestion control and reliability. Globally broadcast, streamed multicast *sessions* are a combination of the standard media types of audio, video, wb, and text. Information about these sessions are periodically transmitted across the MBone on a well-known multicast address. Using the Session DiRectory (sdr) tool[Handley,

1995], users can obtain a list of advertised sessions. Through sdr, a user can choose from this list and launch various multimedia tools. For each of these tools there is a multicast group which the user *joins* when the tool is started. Once part of the group, members will receive group transmissions and they can actively participate or simply listen. Joining a group means that a user must be *grafted* into the multicast tree. The existing multicast routing protocols are capable of seamlessly providing both join (graft) and leave (prune) functions.

## 3. SATELLITE DELIVERY OF MULTICAST

From an IP perspective, a satellite network is essentially a single hop broadcast link. In the case where data is available at the satellite uplink site, the process of transmitting data is straightforward. The more challenging situation is when the receiver is located remote to the uplink site and data must first be transmitted across a terrestrial network. This second situation is described in the next section.

When multicast was first evolving, satellite networks provided an ideal network over which to run one-to-many applications. Because the satellite network represented a single hop, there was no need for a multicast routing protocol. The question of whether to actually send data on a multicast channel could be answered in two ways. First, it could simply be assumed that there would be receivers and so data was always sent. The typical application for this would be a widely popular television station. The second way to answer the question assumes that content for a particular group is less popular. In this situation, satellite network operators often have used a "proxy IGMP" solution. When a host in the satellite footprint wants to join a multicast group, it creates an IGMP message and sends it over the back channel to the uplink site. Upon receiving this message, the uplink site starts transmitting the content over the satellite link.

The end result is that in order to deploy multicast, satellite network providers only needed to deploy a subset of the protocols needed by the terrestrial Internet. This greatly reduced complexity and created an environment where several companies launched successful commercial products.

## 4. INTEGRATING SATELLITE AND TERRESTRIAL NETWORKS

The more challenging problem is how to provide gateway functionality for bridging multicast data delivered from a source in the terrestrial Internet to a satellite network. We demonstrate this through experiments in connections with IETF meetings. IETF meetings have been doing multicast for many years – during each meeting, the audio and video of two selected sessions are to

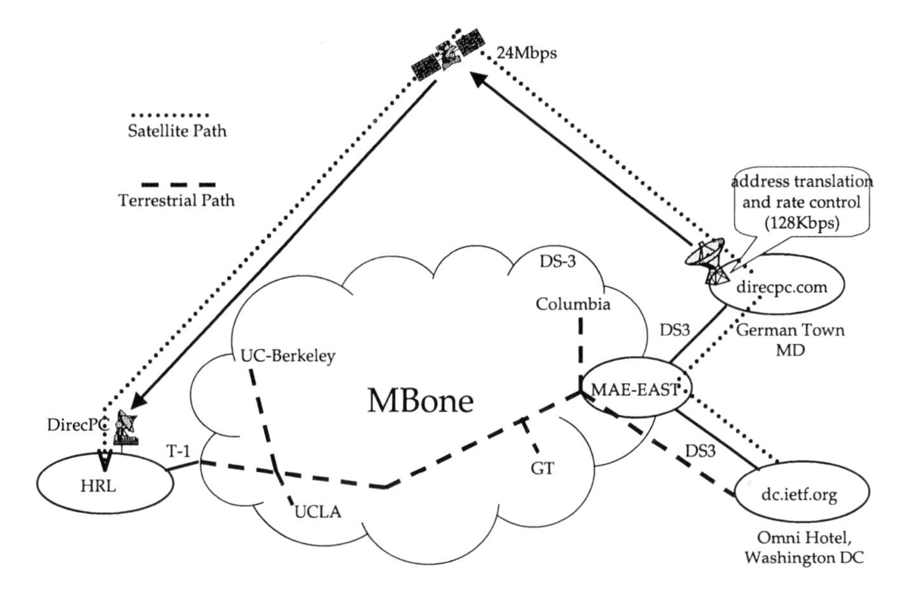


*Figure 5.1.* Logical architecture of the MBone-Over-Satellite experiment during the 40th IETF

multicasted over the MBone. In our experiments, the goal is to bridge these multicast sessions over the satellite network. For the first time at the 40th IETF meeting in Washington DC, during the week of December 8-12, 1997, both IETF multicast sessions were delivered over the satellite [Zhang, 1998]. The primary lessons in the first attempts was how to bridge streaming multicast data onto a satellite network. Then, we evolved this service with our followup effort for the 42nd IETF meeting in Chicago, during the week of August 24-28, 1998 [Almeroth and Zhang, 1998].

Figure 5.1 shows the topology of the multicast-over-satellite experiment. The first part of the experiment took advantage of the fact that IETF sessions are delivered by the IETF staff. This satisfied our requirement for a control group with which to compare the satellite-based results. The two IETF channels were transmitted from the Omni Hotel in Washington DC and connected to the multicast infrastructure via the Mae-East exchange point. We had several remote receivers collecting quality data at their site. We also had one site collecting RTCP feedback from all receivers. The only disadvantage of using RTCP collected at the source is the potential loss of RTCP feedback reports from group members, especially those who are experiencing high loss. However, some percentage of reports will likely be received, even across congested links.

For the satellite component, we used the DirecPC[2] uplink center in German Town, Maryland, to (1) receive the IETF channels; (2) perform address translation; and (3) transmit the sessions over DirecPC. The address translation was necessary to avoid address collision with the terrestrial multicast groups. On the down link side of the satellite there was a DirecPC receiver located at HRL in Malibu, California. It too was collecting data about the quality of the received stream.

# 5. USING SATELLITE PATHS FOR MULTICAST SESSIONS

Quantitatively trying to prove that using satellite paths instead of terrestrial paths offers significant advantages depends on our ability to perform work in three areas. The first is to identify the key performance metrics; the second is to collect and analyze sufficient data to understand performance; and the third is to characterize the situations in which satellite links offer the most advantages. Our efforts in each of these three areas are described in the following sections.

## 5.1 Motivation and Metrics

There are a number of reasons why satellites seem a better choice for the delivery of streaming multimedia data. However, there are also some concerns which may limit the utility of satellite links. First, the potential benefits include:

- **Availability of Bandwidth**: Satellite networks offer a great deal of bandwidth with some systems capable of delivering data at rates of gigabits per second.

- **Lower Loss Characteristics**: In general, a satellite network has much lower loss rates than does the terrestrial Internet, especially when considering the growing congestion in Internet hot spots.

- **Infrastructure Support for Multicast**: When the signal from an orbiting satellite hits the Earth it has a large area of coverage. The broadcast nature of this signal makes the provision of multicast much easier. For example, a signal from a single satellite can span the entire continental United States. A signal intended for one user consumes as much bandwidth as a signal intended to reach thousands of receivers. On a related note, the common approach of providing an asymmetric, terrestrial back channel means traditional multicast protocols do not work the same way. This characteristic is both an advantage and a disadvantage. For example, routing protocols may not be needed given that the satellite network is a one-hop broadcast-to-all medium.

- **Less Dependence on Wireline Infrastructure**: One of the common uses of multicast until now has been for the one-time delivery of conferences, workshops, and meetings. Providing a sufficiently fast Internet connection for these events can be both difficult and costly. One possible scenario is to use a satellite uplink to provide sufficient bandwidth, and then use permanent down links at strategic points to provide connectivity back into the Internet.

However, satellite links do have their limitations. The three that we have experienced in our work are:

- **Increased Delay**: Satellites obviously have an increased propagation delay. The question though is how much, and how do results for satellite delivery compare to terrestrial results.

- **Increased Jitter**: Qualitative results to date suggest that jitter values are also higher for satellite links. There may be a number of factors for this, but our best hypothesis is that the process of taking bits from the terrestrial network and putting them onto the satellite network is adding jitter.

- **Reaching an Uplink Point**: While injecting traffic into the satellite network can potentially be done from anywhere, we have been constrained to using a terrestrial path to get to DirecPC's main uplink site in Germantown, Maryland. This negatively affects performance though we have tried to compensate by removing the terrestrial-to-uplink component in our analysis.

Given our intuition that satellites may be better for certain types of applications, we need to develop a way to quantitatively compare the two. The two sets of metrics we consider are performance and cost. While the focus is on performance, we briefly discuss our preliminary efforts to study resource usage. Performance is measured in terms of the quality and timeliness of the transmission. These components are further refined to include packet loss, delay, and jitter.

The second metric, cost, is more difficult to evaluate. The first problem is there is no good common unit or basis for comparison. Trying to use a common unit like dollar costs is somewhat arbitrary because the process of converting resource usage to dollars is an inexact process. This conversion is straightforward for the satellite path because satellite operators typically charge for a specified amount of bandwidth for a specified duration. However, determining a dollar cost to charge an end user for a terrestrial multicast group is very difficult. Trying to determine the cost of unicast packets or streams is difficult enough without considering the added problem of dealing with multicast transmissions [Chuang and Sirbu, 1998; Herzog et al., 1997; Almeroth and Ammar, 1997]. Our goal is to first study resource usage and then attempt to develop a suitable model for comparison.

## 5.2 Methodology

Data was collected for this research using two tools during the time in which the session was happening, and one additional tool for post-event processing. The types of data collected and the tools used for each include:

- **Group membership using** *mlisten*: The mlisten tool[Almeroth, 1996] was used to collect group membership, join inter-arrival time, and membership duration information for each component of the two IETF channels. The *mlisten* tool works by joining each multicast group and noting the arrival of RTCP packets from each group receiver. This data is archived and can be used to reconstruct a group's membership over time. The *mlisten* tool does not look at data contained in RTCP packets and so does not collect information about packet loss and jitter.

- **Collecting packet loss, jitter, and delay**: All of the transmitted IETF sessions were recorded at the source using the rtpdump utility[Schulzrinne, 1997]. This tool collects both RTP and RTCP packets. These files can be processed and the RTCP feedback packets from all receivers can be used to determine loss, jitter, and an estimate of the round-trip time. The satellite receiver used a similar method, but statistics were gathered in real-time as the session was happening. Similar techniques have been used to study loss in other sessions[Handley, 1997]. The biggest challenge was trying to estimate one-way delay both for terrestrial- and satellite-based receivers. For terrestrial receivers we use RTP sequence times and timing information from RTCP feedback packets to estimate the round-trip time. Because of asymmetries in the satellite network we use a different approach for the DirecPC receiver. Using RTCP-style packets we compute the terrestrial round trip time and the terrestrial-plus-satellite round trip time. We can then estimate one-way delay over the satellite using these numbers.

- **Multicast tree construction**: In order to record the multicast routes used from the source to each receiver, the *mhealth* tool[Makofske and Almeroth, 1998] was used. Mhealth is a management tool that captures RTCP data and performs an *mtrace*[Fenner and et al., 1996] to each receiver. This data was captured during the 42nd IETF and will allow us to generate a representation, over time, of the number of links used in the multicast tree for a particular group of receivers.

## 5.3 Results

In this section we show some of the results for data collected during the 42nd IETF held in Chicago, Illinois, USA, during the dates August 24-28,

1998. The results presented focus either on the entire week, in the case of group membership tracking, or on one particular day, Tuesday, August 25, in the case of loss, jitter, and delay. These results are a subset of the data collected but generally represent behavior commonly observed for other IETF meetings and other multicast sessions. Finally, the satellite-specific results presented in this section were observed over a particular configuration using the DirecPC satellite network. While these results will likely be typical of other satellite networks, without results from other satellite networks we can only suggest that our hypothesis is that results will be similar.

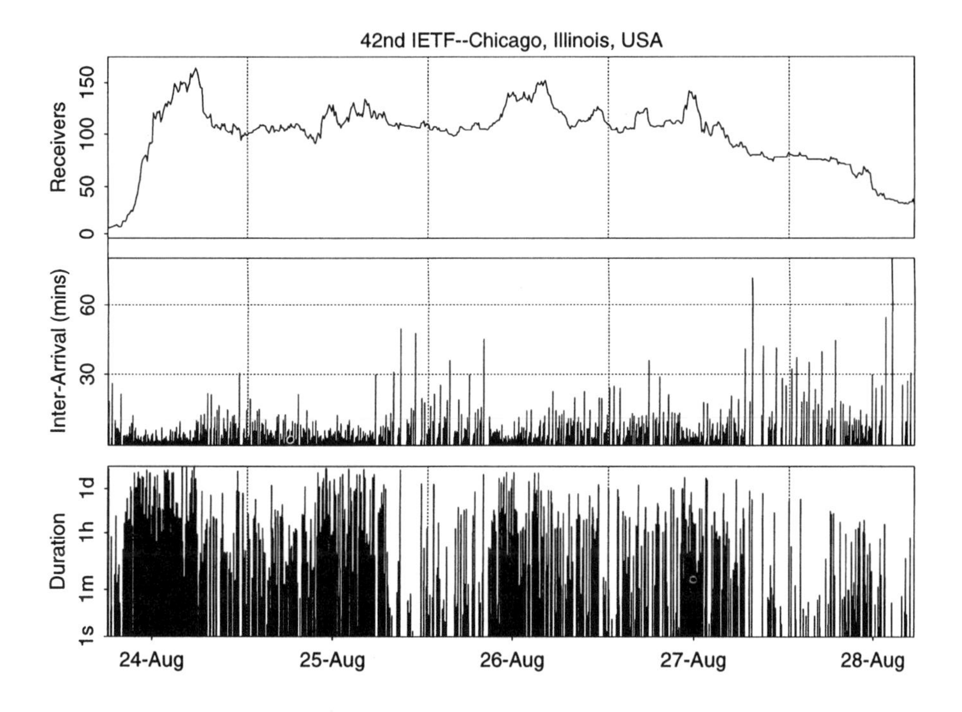


*Figure 5.2.* Group membership details for the 42nd IETF groups

Figure 5.2 shows group size, inter-arrival time, and group membership duration for both IETF channels and all three media formats. The average values for each metric are shown in the graph. Again, results are consistent with past IETFs, i.e. membership and group join/leave activity increases during periods of active transmission. However, it might be worthwhile to note that there is a growing trend for smaller multicast audiences. One generally agreed upon conclusion is that the quality of multicast transmission is not increasing or is even decreasing. Multicast receivers are becoming more dissatisfied with the current quality levels and are less likely to watch multicast sessions. A second interesting observation is that for many multi-day sessions there is no activity at "night" and so membership dips dramatically. However, the IETF has a policy of retransmitting all of a day's sessions at night so receivers on the opposite side of the world can see the sessions during their "daytime". For this reason, the group membership does not decrease as much during the night.

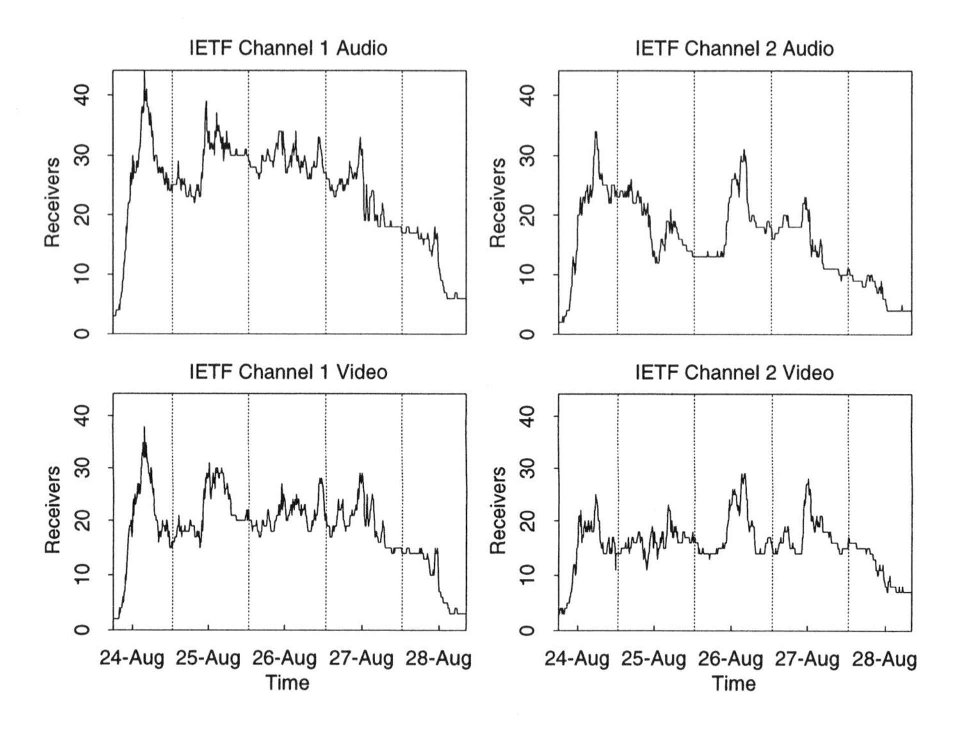


*Figure 5.3.* Breakdown of group membership for each IETF audio/video channel

Figure 5.3 shows a breakdown of group size for each of the audio and video groups for both IETF channels. The results show typical behavior for moderately sized multicast groups[Almeroth and Ammar, 1997; Almeroth and Ammar, 1996]. The data used in Figures 5.2 and 5.3 is important because it can be combined with topology information to re-construct the multicast tree over the course of the session. The data can also be correlated with active periods of transmission and loss patterns. For each group, multicast trees are built by using the minimum set of links to connect a source to all receivers at a particular point in time.

Figures 5.4 and 5.5 show the average loss for each receiver who participated in any of the four sessions on Tuesday, August 25, 1998. Figure 5.4 shows the average loss for the 110 audio receivers and Figure 5.5 shows the loss for the 140 receivers in the video group. The two horizontal lines in each graph represent the average for all terrestrial receivers, and the average for the DirecPC-based receiver (these lines are difficult to see because the average is nearly zero).

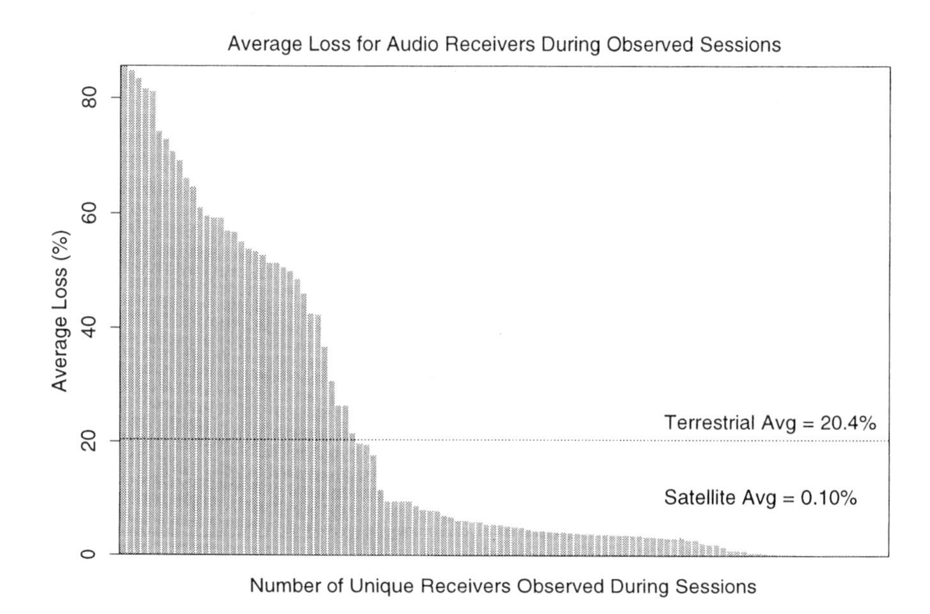


*Figure 5.4.* Packet loss for Channel 1 audio receivers on Tuesday, 25-Aug-98

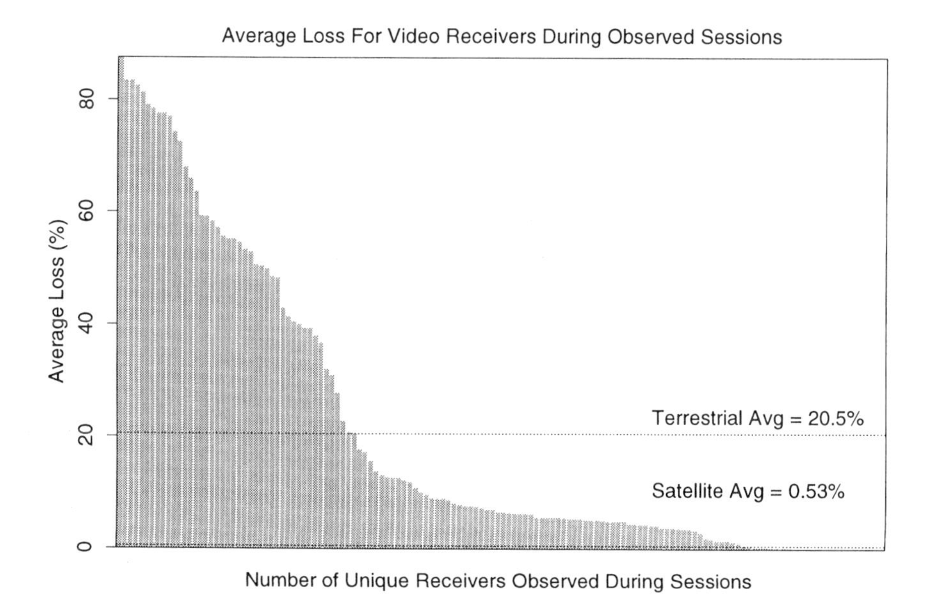


*Figure 5.5.* Packet loss for Channel 1 video receivers on Tuesday, 25-Aug-98

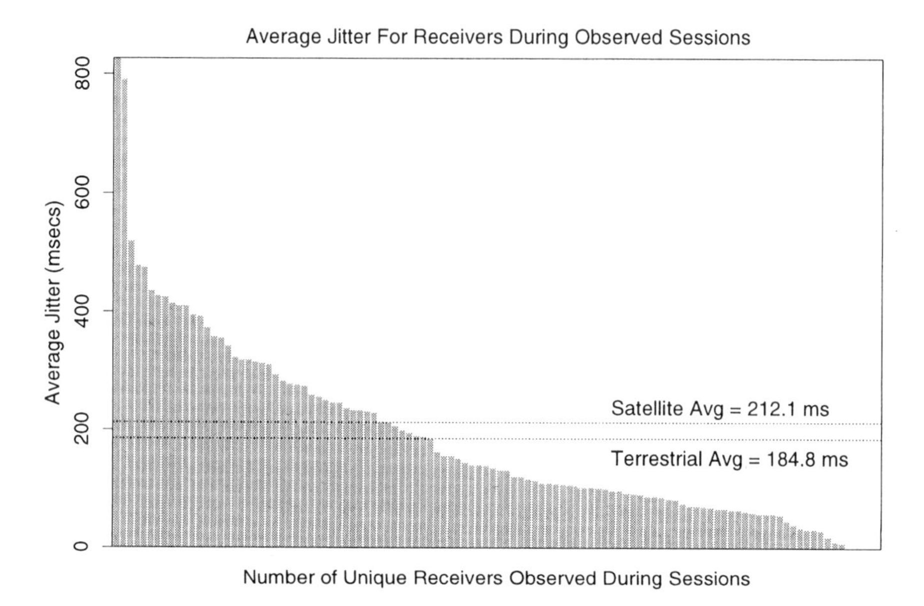


*Figure 5.6.* Jitter for selected terrestrial and DirecPC sites receiving IETF sessions

These results were computed by averaging all of the RTCP packets received from each group member during the Tuesday IETF sessions broadcast on Channel 1. These results are aggregated and averaged over the entire period.

The results show that both audio and video have similar loss characteristics. The DirecPC-based transmission has almost no loss. This can be attributed to the fact that the satellite network can reserve bandwidth and typically has very little congestion. In the case of the terrestrial Internet, more than half of the receivers have less than 20% loss but the few receivers with very high loss arbitrarily increase the average dramatically. Trace results during the IETF suggest that the primary reason was that the multicast tree included several very heavily congested links, including trans-Atlantic and trans-Pacific links.

Figure 5.6 shows the jitter results for the same set of sessions and same set of receivers as was used for the audio loss results shown in Figure 5.4. The key result is that jitter via the DirecPC link is only slightly higher than the average for all the terrestrial-based receivers. This shows, at least for this set of experiments, that jitter is not significantly worse than for the terrestrial network. However, while jitter for the DirecPC receiver is not significantly different than the average, it is still worse than 65% of terrestrial receivers. This suggests there is still a need to investigate the source of jitter in the DirecPC network and attempt to reduce it. In addition, techniques can be used to dampen the affect

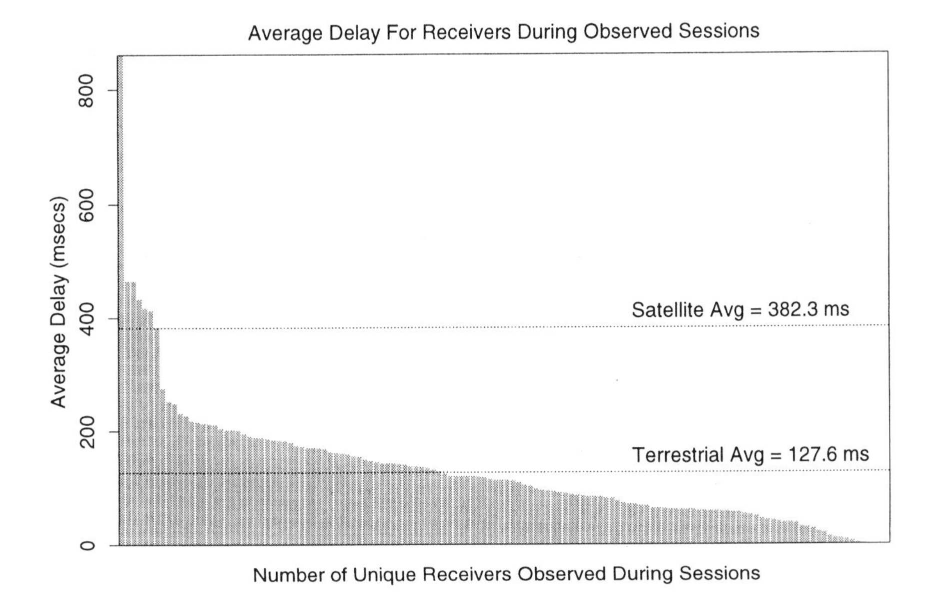


*Figure 5.7.* One-way delay for selected terrestrial and satellite sites

of the jitter. For example, a small amount of buffering and delay can be used at the receiver to wait for delayed packets. However, additional buffering adds additional delay, which for real-time, interactive sessions can be a drawback.

Figure 5.7 shows an estimate of the average one-way delay experienced by each video receiver to participate in the Tuesday IETF sessions transmitted on Channel 1. The general conclusion is that the delay for the DirecPC-based receiver is more than 2.5 times the average of the terrestrial-based receivers. This result was expected given the long propagation delay of satellite networks. The comparison would have been even worse if not for several terrestrial receivers who had very long delays. Further investigation into this set of receivers shows that all of them where located in Europe and had to cross very congested trans-Atlantic links. This set of hosts is also responsible for some of the highest audio and video losses, and some of the largest jitter measurements.

These results were computed using a combination of RTCP statistics and ping packets. For the terrestrial receivers, RTCP provides a mechanism to estimate round trip time because the source sends its timestamp in each RTP packet, and this is returned in the RTCP packet plus additional information about the elapsed time between when the receiver got the packet and when it sent the RTCP feedback packet[Schulzrinne et al., 1996]. Computing an estimate of one-way delay is a matter of computing the round trip time, subtracting the receiver's "hold" time and dividing by two. Computing the one-way delay for the DirecPC receiver was harder because of the asymmetric delay inherent in using the satellite network in one direction and a terrestrial path for the reverse direction. The method used was to send two RTCP-style packets from the DirecPC receiver to the DirecPC uplink site; one response was returned via the satellite link and the other was returned via the terrestrial Internet. By subtracting half of the terrestrial round trip time from the terrestrial/DirecPC round trip time we can estimate the DirecPC-only, one-way delay.

## 6. WHEN TO USE SATELLITES?

Investigation into the use of satellite links is ongoing, but results observed over DirecPC suggest quality can be significantly improved with only a slight increase in jitter and a fair increase in delay. These results suggest that using satellite links to deliver multicast transmissions makes the most sense for sessions with the following characteristics:

- **Large, disperse groups**: Terrestrial multicast is most efficient when groups are not widely scattered. Because of the inherent broadcast nature of satellites, each additional receiver can receive data for no additional cost, no matter how dispersed group members are. The issue of how to best provide inter-continental satellite connectivity is still an open issue.

- **Avoiding terrestrial hot spots**: Satellites offer performance gains for receivers whose terrestrial routing would carry them over links with high packet loss due to congestion. For example, the packet loss percentages to some sites makes the terrestrial multicast infrastructure unusable but acceptable quality can be achieved using a satellite link. Furthermore, satellite down links do not necessarily need to terminate at a end host. A down link might be placed in a strategic location somewhere in the backbone allowing a transmission to bypass a congested network access or exchange point. The decision on where to put down links is an issue left for future work.

- **Pre-arranged broadcast sessions**: Because of the unidirectional nature of most satellite systems, the most effective use of satellite links is as a unidirectional broadcast. Other types of sessions, like truly interactive collaborations, will be more difficult because of the additional delay and jitter. However, VoD-style sessions or planned program broadcasts that do not have as strict timing requirements would be more suitable for delivery via satellite links. And finally, a session that has well known start and stop times would make reservation and assignment of bandwidth in the satellite network easier.

## ACKNOWLEDGMENTS

We thank Hughes Network Systems and its DirecPC division for providing satellite bandwidth and technology assistance. We would also like to express our gratitude to the DirecPC NOC personnel, especially Matt Kenyon, Richard Lodwig, and Bill Donnellan for their endless helps. Finally, we would like to thank David Makofske for assisting in the collection of data at the 42nd IETF.

An early version of this chapter appears in the Third ACM International Workshop on Satellite-based Information Services [Almeroth and Zhang, 1998].

## Notes

1. The standardization body that sets the standards for the Internet (see `http://www.ietf.org/`).

2. DirecPC is a satellite network that delivers Internet traffic over a GEO satellite to anywhere in the United States. See http://www.direcpc.com for more information.

## References

Almeroth, K. (1996). *Multicast Group Membership Collection Tool (mlisten)*. Georgia Institute of Technology. URL: http://www.cc.gatech.edu/computing /Telecomm/mbone/.

Almeroth, K. (2000). The evolution of multicast: From the MBone to inter-domain multicast to Internet2 deployment. *IEEE Network*.

Almeroth, K. and Ammar, M. (1996). Collection and modeling of the join/leave behavior of multicast group members in the MBone (extended abstract). In *HPDC '96*, pages 209–216, New York, USA.

Almeroth, K. and Ammar, M. (1997). Multicast group behavior in the Internet's multicast backbone (MBone). *IEEE Communications*, 35(6):224–229.

Almeroth, K. and Zhang, Y. (1998). Using satellite links as delivery paths in the multicast backbone (mbone). In *Proceedings of the Third ACM International Workshop on Satellite-based Information Services*, Dallas, Texas.

Ballardie, T., Francis, P., and Crowcroft, J. (1995). Core based trees (CBT): An architecture for scalable multicast routing. In *ACM Sigcomm 93*, pages 85–95, California, USA.

Bates, T., Chandra, R., Katz, D., and Rekhter, Y. (1998). Multiprotocol extensions for BGP-4. Internet Engineering Task Force (IETF), RFC 2283.

Casner, S. (1994). *Frequently Asked Questions(FAQ) on the Multicast Backbone(MBone)*. USC/ISI. Available from ftp://ftp.isi.edu/mbone/faq.txt.

Casner, S. and Deering, S. (1992). First IETF Internet audiocast. *ACM Computer Communication Review*, pages 92–97.

Chuang, J. and Sirbu, M. (1998). Pricing multicast communication: A cost based approach. In *INET '98*, Geneva, Switzerland.

Deering, S. and Cheriton, D. (1990). Multicast routing in datagram internetworks and extended LANs. *ACM Transactions on Computer Systems*, pages 85–111.

Deering, S., Estrin, D., Farinacci, D., Jacobson, V., Liu, G., and Wei, L. (1996). PIM architecture for wide-area multicast routing. *IEEE/ACM Transactions on Networking*, pages 153–162.

Farinacci, D., Rekhter, Y., Lothberg, P., Kilmer, H., and Hall, J. (1998). Multicast source discovery protocol (MSDP). Internet Engineering Task Force (IETF), draft-farinacci-msdp-*.txt.

Fenner, B. and et al. (1996). *Multicast Traceroute (mtrace) 5.1*. Available from ftp://ftp.parc.xerox.com/pub/net-research/ipmulti/.

Handley, M. (1995). *SDR: Session Directory Tool*. University College London. Available from ftp://cs.ucl.ac.uk/mice/sdr/.

Handley, M. (1997). An examination of MBone performance. Technical Report ISI/RR-97-450, University of Southern California (USC), Information Sciences Institute (ISI). Available from ftp://north.east.isi.edu/~mjh/mbone.ps.

Herzog, S., Shenker, S., and Estrin, D. (1997). Sharing the 'cost' of multicast trees: An axiomatic analysis. *IEEE/ACM Transactions on Networking*, 5(6):847–860.

Makofske, D. and Almeroth, K. (1998). *Mhealth – Real-Time Multicast Tree Health Monitoring Tool*. Available from http://imj.ucsb.edu/mhealth/.

Rajvaidya, P. and Almeroth, K. (2001). A router-based technique for monitoring the next-generation of internet multicast protocols. In *International Conference on Parallel Processing*, Valencia, SPAIN.

Schulzrinne, H. (1997). *README for RTPtools*. Columbia University. Available from ftp://ftp.cs.columbia.edu/pub/schulzrinne/rtptools/.

Schulzrinne, H., Casner, S., Frederick, R., and V., J. (1996). RTP: A transport protocol for real-time applications. Technical Report RFC 1889, Internet Engineering Task Force.

Zhang, Y. (1998). Integrating satellite networks with internet multicast backbone (mbone over satellite). In *NASA Workshop on Satellite Networks: Architectures, Applications, and Technologies*, Cleveland, OH, USA.

Chapter 6

# TCP PERFORMANCE OVER SATELLITE CHANNELS

Thomas R. Henderson
*Boeing Phantom Works*

**Abstract** The Transmission Control Protocol (TCP) is an integral part of many popular Internet applications, including email, file transfer, and web browsing. Historically, the performance of TCP over satellite channels has been suboptimal due to a variety of protocol algorithm and configuration issues. In this chapter, we describe the challenges that the satellite environment poses to TCP performance. We then summarize a number of standard TCP options that, when applied to a connection, can improve performance. Finally, we survey additional proposals for further improvements that are still considered to be in the research phase.

**Keywords:** TCP, Asymmetry, Congestion Control, Error Recovery, Fairness

## 1. INTRODUCTION

The Internet is a *best effort* network, which means that packets are neither guaranteed to arrive at the intended destination at all, nor guaranteed to arrive at the destination in the order that they were sent. This fundamental design feature of the Internet has allowed it to scale well, because reliability is implemented at the end-hosts and not within the network. To provide applications with a guaranteed, in-order, data delivery service, a reliable transport protocol can be used over this unreliable network. Many of the most popular Internet applications, such as web browsers, file transfer, electronic mail, and remote terminal access, rely on end-to-end reliability between hosts. Almost all of this traffic uses one dominant transport protocol; namely, the Transmission Control Protocol (TCP). In this section, we first describe the basics of TCP operation, focusing on those aspects that are most relevant to satellite links. Next, we survey the large body of work that has aimed at improving TCP performance over satellite links and other network paths that exhibit characteristics similar to satellite links. Finally, we discuss work on other Internet-related reliable transport protocols.

# 2. TRANSMISSION CONTROL PROTOCOL (TCP) OVERVIEW

TCP was originally specified and implemented for the ARPANET in the 1970s; the original Internet RFC was written in 1981 but was derived from several earlier ARPANET specifications [Postel, 1981]. For a comprehensive overview of TCP operation, the interested reader is directed to [Stevens, 1994]. Over the years, a large number of reliable transport protocols have been invented, but TCP is by far the most prevalent protocol used in today's Internet for reliable unicast transport service.

## 2.1 Basic TCP Operation

TCP provides a reliable, end-to-end, byte-streaming data service (with guaranteed in-order delivery) to applications. A transmitting TCP accepts data from an application in arbitrarily-sized chunks and packages it in variable-length segments for transmission within IP datagrams, with each byte of data indexed by a sequence number. The TCP receiver responds to the successful reception of data by returning an acknowledgment to the sender, and by delivering the data to the receiving application; the TCP transmitter can use these acknowledgments to determine if any data requires retransmission. If on the sending side the connection closes normally, the sending application can be almost certain that the peer receiving application successfully received all of the data. Data transfer can occur bidirectionally, and acknowledgments in one direction can be piggybacked on data segments in the other direction.

TCP is typically implemented in the operating system kernel, and accessed through an Applications Programming Interface (API). The most well known and used API is known as *sockets*, and it provides user-level programs with access to network services like TCP through standard system calls [Stevens, 1994].

## 2.2 Connection Establishment and Release

TCP exchanges specially flagged segments to establish, release, and reset a connection. Three segments are typically required to establish a TCP session: the connection initiator (typically called the *client*) first sends a SYN segment to the connection responder (typically called the *server*), the server responds with an acknowledgment (ACK) of the client's SYN concatenated with its own SYN, and the client then sends back an ACK of the second SYN. To close a connection, both sides send a FIN segment to each other, and respond with an ACK of the FIN. Figure 6.1 shows an example of a TCP connection. A minimum of seven packets are usually required to transfer any data.

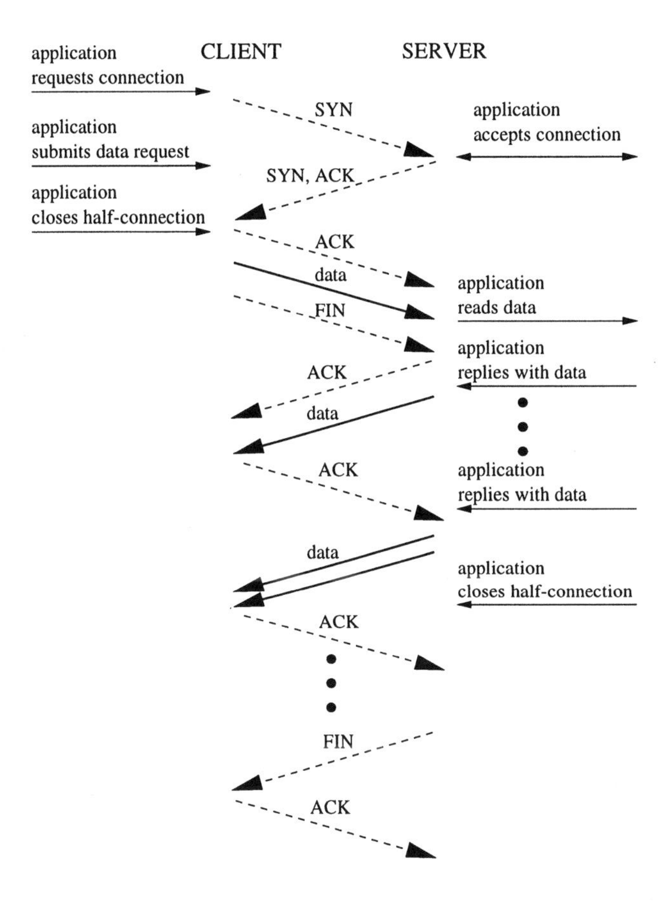


*Figure 6.1.* An example of a TCP data transfer

From an application view, connection establishment may require steps in addition to the SYN exchange described above. These extra steps can contribute to latency in the long RTT satellite environment. One such transaction is the domain name resolution if the peer's IP address is not known. A second transaction may be necessary to establish security relationships with the peer (and network infrastructure such as proxies) if IP security protocols are used. A third possible interaction might be required if an underlying reliable link layer protocol is used. While such interactions at the start of a connection have minor impact to terrestrial connections, their combination can lead to long setup latencies in the satellite environment. Later in this chapter we will introduce an extension known as TCP for Transactions (T/TCP) and show how it can reduce the connection setup latency due to TCP.

## 2.3 Basic Loss Recovery and Congestion Avoidance

TCP's loss recovery and congestion control algorithms are intertwined. The detection of a packet loss causes the TCP sender to retransmit the presumed missing segment, but it also causes the sender to slow down its rate of transmission. The complicated interaction of these algorithms manifests itself significantly in the satellite environment, and we will show below how subtle changes in the algorithms can have substantial impact on performance.

The basic loss recovery mechanism for TCP is a retransmission timer located at the sending end. After a TCP sender sends data, it waits for a *timeout interval* for the receiver to ACK the data. If no ACK is received by the end of the timeout interval, the data is retransmitted and a new timer is started based on a new timeout interval. In most implementations, not every segment is timed– there is only one outstanding segment being timed at any given time. The timeout interval for a segment is based on the estimated round-trip time (RTT) of the connection, and subsequent timeout intervals for the same segment are doubled each time; this process is known as *exponential backoff* of the retransmission timer. The estimated RTT of a connection is obtained by repeatedly timing packet exchanges to obtain RTT samples and subsequently passing the samples through an exponentially weighted moving average filter to obtain a *smoothed* round trip time (srtt) estimate. The initial RTT is assumed to be very large (greater than one second) or may be obtained via a cache. The RTT measurement is usually very coarse in current implementations, and the timeout interval is also very conservative, with the base timeout interval usually set to $srtt + 4 * rttvar$, where $rttvar$ is the mean linear deviation of the RTT measurements.

TCP presumes that packet losses are due to congestion-related buffer overflows in the network, and consequently the detection of a lost packet affects the allowed sending rate of the TCP sender. This algorithm is known as "congestion avoidance." A large part of TCP's success is due to its ability to probe for unused network bandwidth while also backing off its sending rate upon detection of congestion in the network [Jacobson, 1988]. An additional mechanism known as "slow start" is used upon the start of the connection to more rapidly probe for unused bandwidth. The operation of these mechanisms is described in detail in [Stevens, 1994], and is briefly summarized here. TCP maintains a variable known as its *congestion window*, which is initialized to a value of one segment upon connection startup. The window represents the amount of data that may be outstanding at any one time, which effectively determines the TCP sending rate. During slow start, the value of the congestion window is permitted to double every round trip time (RTT), until either a threshold is reached (*slow start threshold*, initially set to an arbitrarily large value), or a loss is detected. All losses are interpreted as congestion events in TCP, so upon a timeout, the

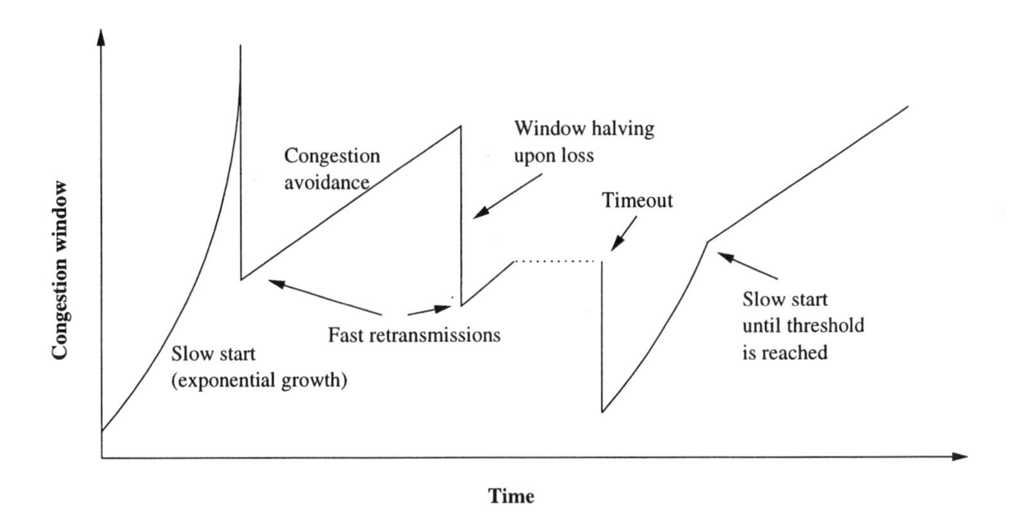


*Figure 6.2.* Basic operation of TCP Reno slow start and congestion avoidance (from [Balakrishnan, 1998])

slow start threshold is set to the current value of the congestion window, the congestion window is then reset to one segment, and TCP begins to slow start again after retransmitting the missing segment. When the window size grows larger than the slow start threshold, TCP enters the congestion avoidance phase, where it adds approximately one segment to its window every one or two RTTs. This is a much slower, linear growth phase of the congestion window.

Slow start and congestion avoidance were introduced into TCP in the late 1980s; TCP implementations that implement slow start and congestion avoidance with the basic loss recovery mechanism described above are known as TCP "Tahoe" implementations.

## 2.4 Enhanced Loss Recovery and Congestion Avoidance

The basic loss recovery described above was the only loss recovery mechanism implemented in the TCP (Tahoe) releases of the late 1980s. An enhancement to TCP Tahoe was added around 1990 to form TCP "Reno." Note that in TCP Tahoe, each time a loss occurs, TCP must wait for a timeout to retransmit the missing segment. Because the timeout interval is conservative, the TCP sender ends up idling for a relatively long period of time (on the order of one to two seconds). Furthermore, the connection must reenter slow start each time a loss occurs. For satellite connections especially, this timeout period and the following slow start result in several seconds during which the throughput is very low and channel bandwidth may be wasted. TCP Reno introduced the "fast retransmit" and "fast recovery" mechanism. TCP Reno assumes that the arrival of three or more duplicate ACKs is a good indication that the segment beyond that which is being ACKed has been lost. Rather than wait for a timeout, it retransmits the segment immediately, and reduces the congestion window to half of its previous value. It then allows the TCP sender to send a new segment for each duplicate ACK received, to keep the transmission pipe full during this recovery phase. If the retransmission is again lost, TCP must wait for a timeout. For single loss events, TCP Reno is very effective in recovering the loss without a damaging reduction of throughput. TCP Reno is described in more detail in [Stevens, 1997]. Figure 6.2 illustrates an example of how TCP Reno's congestion window evolves over time [Balakrishnan, 1998].

The algorithms described above are conservative in nature, a fact that contributes to the stability of the Internet. However, the cost of this stability is a reduction in peak throughput performance, particularly over paths with long feedback loops such as the long delay of satellite links. As we describe below, a significant amount of effort has gone into enhancing these algorithms for environments such as satellite links, while still maintaining TCP behavior that is considered "friendly" to other competing packet flows.

## 3. TCP PERFORMANCE PROBLEMS OVER SATELLITE LINKS

Satellite networks formed a part of experimental internets beginning in the mid 1970's (in the form of the Atlantic Packet Satellite Network, or "SATNET" [Jacobs et al., 1978]), and TCP is reported to have worked correctly over such links, albeit at bit rates in the tens of Kb/s [Seo et al., 1988]. However, performance problems did not manifest themselves in a network where the maximum link capacity was 56 Kb/s. In this section, we summarize some of the solved and unsolved TCP performance problems in a satellite environment. Partridge and Shepard also discuss several of these causes for poor satellite TCP performance in [Partridge and Shepard, 1997].

The main characteristics of the end-to-end path that affect transport protocol performance are latency, bandwidth, packet loss due to congestion, and losses due to transmission errors. If part of the path includes a satellite channel, these parameters can vary substantially from those found on wired networks.

- **Latency:** The three main components of latency are propagation delay, transmission delay, and queueing delay. In the broadband satellite case, the dominant portion is the propagation delay. For connections traversing GEO links, the one-way propagation delay is typically on the order of 270 ms at mid-latitudes, and may be more depending on the presence of bit interleavers for forward error correction. Variations in propagation delay for GEO links are usually removed by Doppler buffers. In the LEO case, the absolute delay can be an order of magnitude less. For example, satellites at an altitude of 1000 km will contribute roughly an additional 20 ms to the one way delay for a single hop; additional satellite hops will add to the latency depending upon how far apart are the satellites. However, the delay will be more variable for LEO connections since, due to the relative motion of the LEO satellites, propagation delays will vary over time, and the connection path may change. Therefore, for LEO-based transport connections, the propagation delay will generally be smaller (such as from 20-200 ms), but there may be substantial delay variation added due to satellite motion or routing changes, and the queueing delays may be more significant [Gavish and Kalvenes, 1998]. The previous chapter provides illustrations of the type of delay variability that may be encountered.

- **Asymmetry:** With respect to transport protocols, a network exhibits asymmetry when the forward throughput achievable depends not only on the link characteristics and traffic levels in the forward path but also on those of the reverse path [Balakrishnan et al., 1997]. Satellite networks can be asymmetric in several ways. Some satellite networks are inherently bandwidth asymmetric, such as those based on a direct broadcast satellite (DBS) downlink and a return via a dial-up modem line. Depending on the routing, this may also be the case in future hybrid GEO/LEO systems; for example, a DBS downlink with a return link via the LEO system causes both bandwidth and latency asymmetry. For purely GEO or LEO systems, bandwidth asymmetries may exist for many users due to economic factors. For example, many proposed systems will offer users with small terminals the capability to download at tens of Mb/s but, due to uplink carrier sizing, will not allow uplinks at rates faster than several hundred Kb/s or a few Mb/s unless a larger terminal is purchased.

- **Transmission errors:** Bit error ratios (BER) using legacy equipment and many existing transponders have been poor by data communications standards; as low as $10^{-7}$ on average and $10^{-4}$ worst case. This is primarily because such existing systems were optimized for analog voice and video services. New modulation and coding techniques, along with higher powered satellites, should help to make normal bit error rates very low (such as $10^{-10}$) for GEO systems. For LEO systems, multipath and shadowing may contribute to a more variable BER, but in general those systems are also expected to be engineered for "fiber-like" quality most of the time (with advances in error correction, links are more likely to be in one of two states: error free, or completely unavailable).

- **Congestion:** With the use of very high frequency, high bandwidth radio or optical intersatellite communications links, the bottleneck links in satellite constellations will likely be the links between the earth and satellites. These links will be fundamentally limited by the uplink/downlink spectrum, so as a result, the internal satellite network may be free of heavy congestion. However, the gateways between the satellite subnetwork and the Internet could become congested more easily, particularly if admission controls were loose.

A variety of solutions to these problems have been proposed. Several solutions involve taking advantage of standardized TCP protocol features, as well as approaches for improving the underlying communications channel performance; these solutions are non-controversial and are advocated by the Internet Engineering Task Force (IETF) as "best practices." We describe these techniques in the next section.

## 4. ENHANCING TCP PERFORMANCE USING STANDARD MECHANISMS

Over the past decade, a number of TCP extensions have been specified which improve upon the performance of the basic protocol in such environments. Many of these approaches are recommended by IETF RFC 2760 [Allman et al., 1999]. Even though some of these options have been specified for over five years, it has only been recently that implementations commonly support them. The lack of widespread vendor support for satellite-friendly protocol options has historically been a hindrance to achieving high performance over satellite networks. Nevertheless, since operating systems continue to be under rapid development, the barriers to deployment for TCP enhancements are relatively lower than for other protocols.

### 4.1 Window scale

TCP's protocol syntax originally only allowed for windows of 64 KB, which limited throughput in practice to a theoretical 900 Kb/s. The window scale option [Jacobson et al., 1992] significantly increases the amount of data which can be outstanding on a connection by introducing a scaling factor to be applied to the window field. This is particularly important in the case of satellite links, which require large windows to realize their high data rates. Because of window scale, researchers have reported TCP throughputs over geostationary satellite links (in controlled environments with no congestion or bit errors) in excess of 100 Mb/s [Charalambous et al., 1998]. The window scale option should be used in conjunction with two companion algorithms known as "Protection Against Wrapped Sequence Space" (PAWS) and "Round-Trip Time Measurements"

(RTTM) [Allman et al., 1999], of which the latter slightly increases the TCP header overhead.

## 4.2 Path MTU discovery

For satellite-based TCP connections, the practice of sending the largest segments possible has two main benefits: i) it reduces the fraction of bandwidth consumed by header overhead, and ii) it allows the transmission of the maximal amount of data possible under the constraints of the congestion window. However, if a segment size that is too large for the path is selected, costly fragmentation and reassembly will occur at the IP layer somewhere along the path.

The Path MTU discovery option [Mogul and Deering, 1990] allows a TCP connection to explore a network path for the largest allowable Message Transfer Unit (MTU). In this mechanism, a sending TCP implementation will probe the network path by sending a segment of the appropriate size for the network segment to which it is connected (e.g., 1500 bytes for Ethernet), with the "don't fragment" bit set in the IP header. If this segment is too large for the path, a downstream router will generate an ICMP message to the TCP sender, and the sender will try again with a smaller segment size (identified in the ICMP message).

One disadvantage of Path MTU discovery is that the process of discovering the proper MTU incurs latency at the beginning of a connection. This may be partially alleviated by allowing implementations to cache recently used MTUs, although the proper policies for aging the cache values (to adjust for the possibility of dynamic routing changing the path) is still an open question. Another potential problem is that the use of large segments may conflict with optimal segment sizes on a satellite channel with a higher bit error ratio. Despite these tradeoffs, the IETF recommends the use of Path MTU discovery over satellite channels [Allman et al., 1999].

## 4.3 Error correction

TCP treats all losses as a sign of congestion. If a segment is lost to a transmission error, TCP misinterprets the loss as congestion and inappropriately responds by reducing the congestion window. Unnecessary reductions of the congestion window are particularly damaging to throughput over satellite channels. The error rate on satellite channels is a result of the design of the link, and since analog voice channels have been relatively forgiving of high bit error ratios (BER), many legacy satellite links were not engineered to provide low BERs (in order to save cost). However, even mild error rates on very high speed satellite links can have a crippling effect on TCP throughput [Charalambous et al., 1999].

The practice of employing stronger forward error correction (FEC) channel coding on satellite links can help here. Recent advances in concatenated error control codes can make most broadband satellite channels relatively error free. The use of additional coding at the physical layer does have the drawback of requiring additional overhead, as well as additional hardware at the sending and receiving sides of a link. However, these costs are more than offset by the performance benefits reaped by data protocols.

Link layer retransmission protocols (historically called "automatic repeat request" (ARQ) protocols) are another possible solution to bit errors. An ARQ protocol will try to recover lost frames and present the IP layer with an error-free sequence of packets. This practice, however, can lead to enhanced packet jitter (because link layer frames must be buffered on the receive side while link retransmission takes place) and could potentially lead to coarse timeouts at the TCP layer if the link layer recovery takes too long. An example of this approach can be found in [Stadler, 1997].

## 4.4 Further loss recovery enhancements

We conclude this section with a more careful look at the impact of newer loss recovery mechanisms on satellite TCP performance. Specifically, options known as "NewReno" [Floyd and Henderson, 1999] and Selective Acknowledgments (SACK) [Mathis et al., 1996] can be very beneficial for long file transfer performance, especially when used in combination.

For all of its effectiveness at recovering from single loss events, TCP Reno has what is widely considered a bug when it comes to multiple loss events in a single window [Fall and Floyd, 1996; Hoe, 1996]. The problem is that if multiple losses occur in a window of data (i.e., within the same RTT interval), TCP Reno only performs fast retransmit of the first missing segment, and often must wait for a timeout for subsequent lost segments. TCP implementations that fix this bug are known as TCP "NewReno." RFC 2582 describes a number of related fixes for this problem [Floyd and Henderson, 1999].

TCP Reno and NewReno can only recover from one loss event every RTT. In an environment where the RTT is large, this leads to a very slow recovery for bursty loss events. TCP with Selective Acknowledgments [Mathis et al., 1996], also known as TCP SACK, standardizes a new TCP option that allows the receiver to report a large number of missing segments at one time. The TCP sender can thereby learn about and recover from multiple missing segments in a reduced number of round trips.

Initial implementations of TCP with Selective Acknowledgments were based on underlying TCP Reno implementations. TCP SACK can also be implemented in conjunction with TCP NewReno extensions. Experimental results suggest that the combination of TCP SACK with TCP NewReno is important

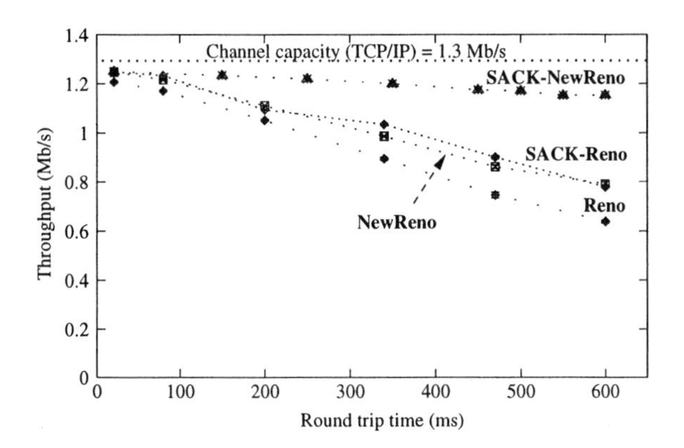


*Figure 6.3.* Experimental file transfer performance of TCP SACK NewReno, TCP SACK Reno, TCP NewReno, and TCP Reno implementations over an emulated satellite path with a TCP/IP bandwidth of 1.3 Mb/s and no transmission errors (from [Henderson and Katz, 1999])

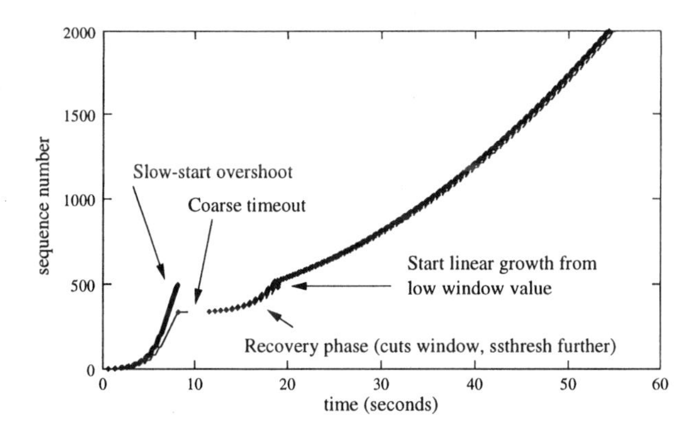


*Figure 6.4.* Typical performance, using a standard BSD TCP (Reno) implementation, of a large file transfer over a GEO satellite channel

in the satellite environment. Figure 6.3 (from [Henderson and Katz, 1999]) illustrates the average throughput performance of 10 MB file transfers over a satellite channel using four TCP variants. Note first that the difference in performance does not manifest itself until round trip times become large. Note also that while both TCP SACK and TCP NewReno can offer benefits over TCP Reno, the performance comes closest to achieving full channel utilization only when both options are combined.

An analysis of packet traces reveals that the main difference between the implementations is in their behavior immediately upon leaving the slow start

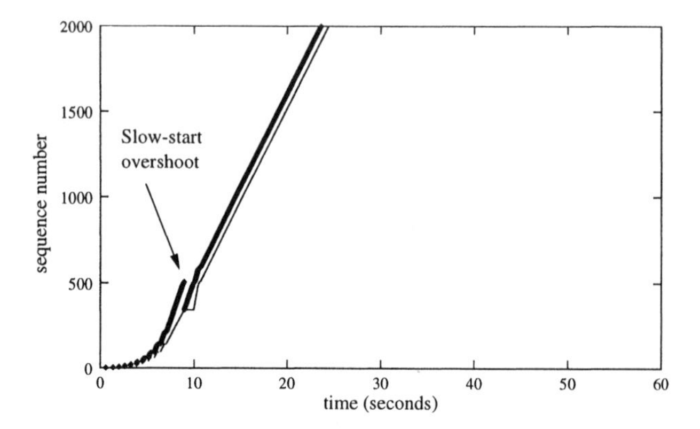


*Figure 6.5.* Correct SACK behavior, using a modified BSD TCP implementation (including NewReno loss recovery), of a large file transfer over a GEO satellite channel

phase of congestion avoidance. For good file transfer performance, it is critical that TCP smoothly transition from slow start to congestion avoidance with a congestion window close to the bandwidth-delay product of the path. Figure 6.4 illustrates a typical packet trace for TCP Reno from the experiment shown in Figure 6.3. TCP Reno rarely avoids a retransmission timeout and multiple reductions in its window after the first slow start, resulting again in slow window growth. The performance of TCP NewReno is only slightly better, because while TCP NewReno avoids multiple reductions in its sending window due to the burst loss, it can only recover one loss per RTT, and it therefore spends a large amount of time performing the actual recovery. The performance of TCP SACK combined with Reno is similar, but for a different reason. In this case, the recovery is rapid, since the SACK option gives the sender a more immediate picture of the losses incurred on the link, but the sending window is repeatedly reduced (since each packet gap in the burst is treated as a "new" loss) and the sender restarts the connection in linear window growth mode from a very low value. Finally, in the case of TCP SACK with NewReno extensions, when a slow start overshoot occurs, the protocol cuts its window in half only once and smoothly moves to congestion avoidance after recovering all losses. There is little penalty for using a high-bandwidth, high-latency satellite link in this case. As an example, Figure 6.5 can be contrasted with Figure 6.4. Further analysis of this behavior is found in [Henderson, 1999].

## 5. RESEARCH ISSUES

Despite the progress on improving TCP's performance, there remain some vexing attributes of the protocol that impair performance over satellite links.

For these problems, there are no standardized solutions, although some are currently under study. In 2000, the IETF issued an informational document that summarizes many of the ongoing research issues [Allman et al., 2000]. In this section, we describe these issues in further detail.

## 5.1 Connection startup

TCP's slow start mechanism, while opening the congestion window at an exponential rate, may still be too slow for broadband connections traversing long RTT links, resulting in low utilization. This problem is exacerbated when slow start terminates prematurely, forcing TCP into the linear window growth phase of congestion avoidance early in the connection [Partridge and Shepard, 1997]. This section describes approaches to speed up the rate of data transfer at the beginning of TCP connections.

Besides file transfers, most of the rest of the TCP traffic in the Internet is driven by web transfers. Such connections are very different from file transfers. Typically, an web client issues a small request to a server for an HTML (HyperText Markup Language) page. The server sends the initial page to the client on this first connection. Thereafter, the client launches a number of TCP connections to fetch images that fill out the requested page or to obtain different pages. Each item on the page requires a separate connection (a more recent modification to this approach, known as Persistent-HTTP (P-HTTP), reuses the same TCP connection for multiple items). Many common web browsers allow a user to operate multiple (typically, four) TCP connections in parallel to fetch different objects. Basically, the data transfer model is "client request, server response."

Using standard TCP, any connection requires a minimum of two RTTs until the client receives the requested data (the first RTT establishes the connection, and the second one is for data transfer). As the RTT increases, the RTT can become the dominant portion of the overall user-perceived latency, particularly since average web server response times are much smaller than one second. Figure 6.6, adapted from a similar figure in [Heidemann et al., 1997], illustrates the latencies associated with a three-segment data exchange.

Two mechanisms can help to alleviate the latency effects of TCP for short connections. The first, T/TCP, does away with the initial handshake (RTT) of the connection. The second, 4KSS, allows the TCP server to send up to 4380 bytes in the initial burst of data. If the size of the transfer is no more than 4380 bytes, the transfer can complete in one RTT.

TCP for Transactions (T/TCP) [Braden, 1994], among other refinements, attempts to reduce the connection handshaking latency for most connections, reducing the user-perceived latency from two RTTs to one RTT for small trans-

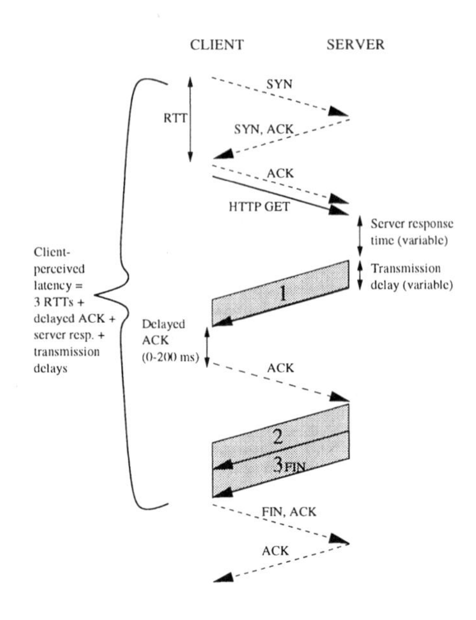


*Figure 6.6.* TCP latency of a 3 segment server reply using standard TCP

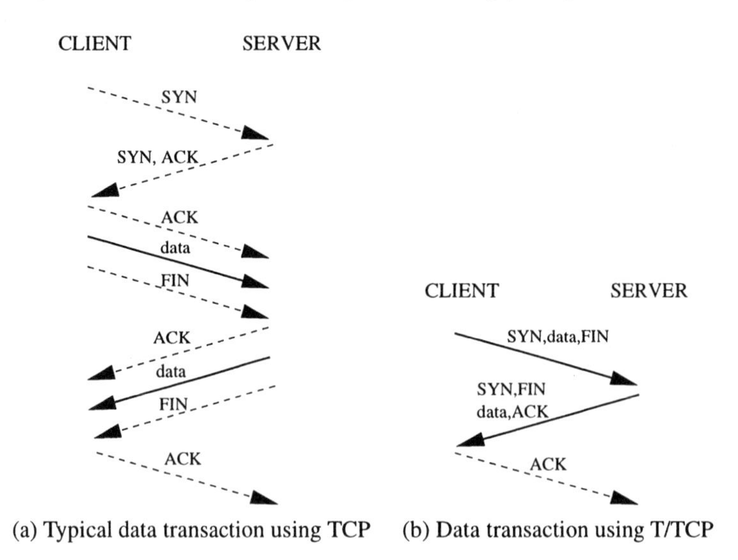


*Figure 6.7.* Typical packet sequences for TCP and T/TCP

actions. This reduction can be significant for short transfers over satellite channels.

Figure 6.7 shows how the minimal seven packet TCP transaction can be reduced to three for a small exchange. Support of the T/TCP extension has been slow, however, for two main reasons. First, there are security concerns over denial of service attacks based on T/TCP (increased vulnerability to SYN floods and attacks on the accelerated open procedure). Second, T/TCP requires the application to use the `sendto()` or `sendmsg()` system calls when instantiating a connection; however, most applications use the `connect()` and then `send()` or `write()` system calls.

Other researchers have considered allowing a TCP connection to use an initial congestion window of 4380 bytes (or a maximum of 4 segments) rather than one segment [Allman et al., 1998]. Transfers for file sizes under roughly four thousand bytes (many web pages are less than this size) would then usually complete in one RTT rather than two or three. In the following, we refer to this policy by its informal name of "4K slow start" (4KSS).

With these mechanisms in mind, consider again Figure 6.6, in which dashed lines denote control packets and solid lines indicate data packets. The first RTT is consumed by a SYN exchange, after which the client issues an HTTP GET request. Upon receiving and responding to this request, the server at this point has a congestion window of one segment. Assuming that the TCP implementation implements delayed acknowledgments (delayed ACKs) of up to 200 ms [Stevens, 1994], the client on average will acknowledge this data after around 100 ms. Upon receiving the acknowledgment, the congestion window grows to 2, and the server sends the second and third segments, followed by a FIN, which closes its half of the connection. The client must close its own half of the connection, but that event does not contribute to user-perceived latency. Therefore, the total amount of TCP-related latency is 3 RTTs plus the delayed acknowledgment delay in this case. Using either T/TCP or 4KSS would reduce the latency to around 2 RTTs, and using both mechanisms would reduce it to around single RTT. Experimental results indicate that latencies for typical transaction can be reduced by over 50% with the use of both mechanisms.

Two other related approaches to speeding up an initial connection involve the delayed acknowledgment mechanism. Since TCP implementations typically prefer to acknowledge every other segment by default, the congestion window opens up roughly half as slow as should technically be permitted (given the conservative assumption that the receipt of an acknowledgment indicates that a packet has left the network in the forward direction). One approach is known as "byte counting," and involves modifying the sender to send new segments based on the amount of data acknowledged, not on the number of acknowledgments received. The second related approach involves receiver-side modifications to inhibit delayed acknowldegments during the sender's slow start (although the means by which the receiver can know when the sender is in slow start are an open issue). These approaches are discussed further in [Allman et al., 2000].

Finally, the HTTP version 1.1 specification recommends that servers and clients adopt the persistent connection and pipelining techniques known as "persistent-HTTP" (P-HTTP) [Padmanabhan and Mogul, 1994]. Rather than using separate TCP connections for each image on a page, P-HTTP allows for a single TCP connection between client and server to be reused for multiple objects. The shift to P-HTTP offers a tradeoff in performance for satellite connections. On the one hand, P-HTTP is potentially much more bit-efficient than HTTP with standard TCP, because connections are not set up and torn down as frequently (the connection establishment costs are identical to those of T/TCP [Heidemann et al., 1997]). However, in terms of latency, the use of T/TCP and multiple, concurrent connections may yield faster web page loads under some scenarios.

## 5.2 Shared TCP state and TCP pacing

Optimal performance of TCP connections results from proper selection of a number of parameters. Using adaptive algorithms such as Path MTU Discovery and congestion avoidance, each TCP connection gradually obtains a set of reasonably good values for these parameters. However, new connections generally start from scratch each time. And while a TCP is probing to discover the best parameters, suboptimal performance is incurred.

Therefore, researchers have investigated the potential for caching congestion information from a recently used connection to start the new connection from a larger initial window size [Padmanabhan and Katz, 1998],[Touch, 1997]. For example, if it is known that a recent connection to a given network was able to obtain a very large window, it may be permissable to assume that a new connection need not start out with a window size of one segment. The sharing mechanism, however, determines priorities among the different connections and can lead to unfair network bandwidth sharing. One research problem involves the proper way to age such cached information, for as time goes on, the conditions in the network assumed by the cached parameters will change. Information may also be shared among many hosts and stored in a server [Seshan et al., 1997], although the question of what particular information should be shared and how to do so securely requires further research. Further details of integrated congestion management issues can be found in [Balakrishnan et al., 1999].

Shared TCP state can also be used for rate-based pacing of TCP segment transmissions. TCP has the property that it uses incoming acknowledgments as a trigger to send new segments. However, were it allowed to send more data initially, perhaps based on shared TCP state or previous activity on a persistent connection, it would not be able to do so using traditional implementations because no ACKs would arrive to trigger the sending of new segments.

Rate-based pacing is a proposed technique to allow a data sender to temporarily pace TCP segments at a given rate to restart the regular ACK flow.

The pacing rate may be known from shared TCP state or previous activity on a persistent connection. Upon receipt of new ACKs, the pacing is discontinued [Allman et al., 2000]. Rate-based pacing could also be used in general to smooth out TCP bursts, which can lead to increased congestion.

## 5.3 Link asymmetry

The throughput of TCP over a given forward path is maximized when the reverse path has ample bandwidth and a low loss rate, because TCP relies on a steady stream of acknowledgments (ACKs) to advance its window and clock out new segments in a smooth manner. When the reverse path has limited bandwidth, the TCP acknowledgment stream becomes burstier, as ACKs are clumped together or dropped. This has three effects: i) the sending pattern also becomes more bursty, ii) the growth of the congestion window (which advances based on the number of ACKs received) slows, and iii) the fast retransmit mechanism that avoids retransmission timeouts becomes less effective. TCP does not have congestion control that operates on the reverse path, and the reverse path queues could potentially get very large, leading to increased latency. Finally, if the MTU for the constrained reverse channel is small, the path MTU discovery mechanism will select the small MTU for the forward path also, reducing performance. The IETF has loosely defined an asymmetric channel as one with a forward to reverse link bandwidth ratio of more than 75:1 (e.g., a 1.5 Mb/s forward path with a 20 Kb/s return path) [Allman et al., 2000].

Since TCP acknowledgments are cumulative, researchers have recently studied ways to reduce the amount of ACK traffic over the bottleneck link by "ACK congestion control." In ACK congestion control, intermediate routers use the Explicit Congestion Notification (ECN) bit once the queue occupancy grows above a given threshold. The sender then is responsible for somehow echoing the ECN information back to the data receiver, which then takes steps to reduce the frequency of ACKs [Balakrishnan et al., 1997]. Since fewer ACKs are received by the sender, this approach must be paired with new mechanisms to grow the window based on the amount of data acknowledged (such as the byte counting strategy studied in [Allman, 1999]) and that more smoothly pace new data transmission by using rate pacing and limits on TCP bursts [Balakrishnan et al., 1997]. Solutions such as this have the drawback of requiring transport-layer implementation changes at both ends of the connection.

An approach that does not require host modifications is known as "ACK filtering and reconstruction" [Balakrishnan et al., 1997; Samaraweera, 1999]. Since TCP acknowledgments are cumulative, a bottleneck router can discard older acknowledgments in favor of the newest one if congestion occurs. ACK reconstruction occurs on the egress side of the slow link, where the missing ACKs are regenerated and spaced corresponding to their original spacing. Although this approach does not require changes at the TCP sender, ACK reconstruction is challenging to implement. ACK filtering without reconstruction could also be used if the sending adaptation mechanisms discussed above were commonly implemented.

A comprehensive study on the use of TCP over asymmetric channels is described in [Balakrishnan et al., 1997], although the motivation for the study was packet radio and wireless cable networks. The authors investigated several techniques for reducing the frequency of ACKs generated by the TCP receiver, by examining both network agent-based solutions that do not require host modifications and solutions involving modifications to the TCP implementation.

## 5.4 Experimental loss recovery techniques

Research continues on additional refinements to TCP's loss recovery mechanisms. Not all refinements are aimed at satellite channels specifically, but at wireless and lossy environments in general. We have already illustrated above how TCP SACK combined with NewReno can yield good file transfer performance over satellite channels. A similar mechanism, developed by Fall and Floyd [Fall and Floyd, 1996], makes use of a new TCP send variable called *pipe*, which estimates the number of segments outstanding in the network. In general, this algorithm allows TCP to recover from multiple segment losses within one RTT by decoupling the decision of when to send a segment from the decision of which segment to send. The IETF allows the use of this algorithm as it is consistent with the overall spirit of fast recovery [Allman et al., 2000].

Another related proposal is the Forward Acknowledgment (FACK) algorithm [Mathis and Mahdavi, 1996]. FACK attempts to decouple congestion control from loss recovery by more sophisticated use of the SACK option, allowing the TCP sender to make more intelligent choices about which segment to send and when. One relevant aspect of the algorithm is that it sends new data segments during fast recovery (in a manner that does not violate the spirit of TCP's congestion control), so as to keep TCPs ACK self-clock running during recovery. This latter feature may be especially beneficial for long delay networks [Allman et al., 2000].

Explicit Congestion Notification (ECN), introduced above, is an active queue management mechanism that allows routers to notify senders of imminent congestion, without dropping segments. This may be implemented by directly sending a control message to the sender ("backward" ECN), or by marking a "congestion experienced" bit in the packet header ("forward" ECN). Experiments have shown that ECN is effective in reducing the overall segment loss rate, which is important in satellite environments because losses require TCP to pause for a RTT or more to recover.

Finally, one of the most challenging approaches is to attempt to distinguish between lost segments and errored segments at the TCP layer. If a segment were merely errored, it could be retransmitted with no reduction in the congestion window. However, there exists no standard way to inform TCP of corruption. One approach proposed uses a new type of ICMP error message ("corruption experienced") generated by routers or gateways along the path– however, the generation of such messages is made more difficult by the fact that the source and destination addresses may themselves be errored. Further, if the path is highly asymmetric, such a "corruption experienced" message might cause further congestion on the reverse path. The use of such mechanisms is not yet encouraged by the IETF, which is concerned about the possibility of ignoring true congestion signals [Allman et al., 2000].

## 5.5 Implementation details

In many implementations, applications must explicitly request large sending and receiving buffer sizes to trigger the use of window scaling options. For example, default socket buffer sizes for many TCP implementations are set to 4 KB [Heidemann, 1997]. Unfortunately, this requires users to manually configure applications and TCP implementations to support large buffer sizes; moreover, some applications and operating systems do not permit such configuration, including common web servers [Heidemann, 1997]. Since larger socket buffers consume more memory, it is not likely that larger socket buffers will be turned on by default. Also, because TCP can only negotiate the use of window scaling during connection setup, unless it has cached the value of the RTT to the destination, it cannot invoke window scaling upon finding out that the connection is a long RTT connection. As we mentioned above, even if T/TCP is present in an implementation, applications based on the sockets Application Programming Interface (API) often use system calls that prevent the usage of T/TCP. Because the TCP standard is not rigorously defined or followed, different vendor implementations often have different (and buggy) behavior (see, for example, [Paxson, 1997] and [Brakmo and Peterson, 1995]). The subtle performance effects of these variations can significantly manifest themselves over satellite channels.

## 5.6 TCP fairness

Perhaps the most challenging problem is that TCP's congestion avoidance algorithm results in drastically unfair bandwidth allocations when multiple connections with different RTTs share a bottleneck link. The bias goes against long RTT connections by a factor of $RTT^{\alpha}$, where $\alpha < 2$ [Lakshman and Madhow, 1997]. This problem has been observed by several researchers beginning with [Hashem, 1989], but a viable solution has not yet been proposed, short of mod-

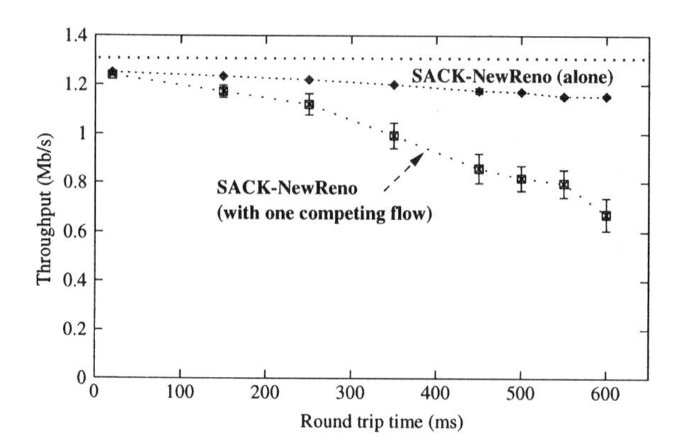


*Figure 6.8.* The effect of a single competing short-delay connection on the satellite connection's throughput (from [Henderson and Katz, 1999])

ifying network routers to isolate and protect competing flows from one another [Suter et al., 1998]. Furthermore, bandwidth asymmetry exacerbates the fairness problems by shutting out certain connections for long periods [Lakshman et al., 1997]. In [Floyd and Jacobson, 1992], the authors discuss a "constant rate" window adjustment algorithm. They observe that RED gateways and Reno-style enhancements to TCP are insufficient to correct the bias inherent in the standard algorithm. In [Floyd, 1991], the performance of a constant rate increase algorithm is evaluated via simulation and qualitative analysis for connections with long RTTs which traverse multiple gateways. The author explores the performance when all connections in the simulation topology employ the modified algorithm, and shows that the performance of the constant rate algorithm meets at least one accepted measure of fairness, while the performance of standard TCP clearly does not. In [Floyd, 1994], Floyd explores the issues surrounding alternative window increase algorithms, and [Henderson et al., 1998] expanded this work by comparing several alternatives in simulation. Finally, Lakshman and Madhow study the performance of TCP/IP in networks with high bandwidth-delay products [Lakshman and Madhow, 1997]. The authors observe that TCP is "grossly unfair" towards connections with higher round-trip delays, and suggest that an alternate dynamic window algorithm is a high priority for future research, although they do not endorse any new algorithm.

We showed above in Section 4.4 how the combination of TCP SACK and NewReno options could yield close to optimal file transfer throughput for a satellite channel. Figure 6.8 illustrates the results of adding a single, large-window persistent connection from a background source to a background sink in the same direction as the foreground file transfer. This background connection does not traverse the satellite link, but shares a common part of the end-to-end path. The throughput results for the satellite link in this case are strikingly different, as the short delay connection is much more aggressive in using the congested link when the long delay connection's RTT is increased. In fact, TCP's fairness properties can be the first-order determinant of how well a large-window satellite TCP connection can do in the wide-area Internet. Even though the satellite connection was successful in avoiding timeouts in almost all of the transfers, the window reductions due to recurring fast retransmits substantially reduced the throughput. The throughput is also much more variable under these conditions, as represented by the error bars. The main problem is that the connection with the long RTT is too sluggish to rebuild its window and push data through the congested queue before it takes another loss.

We close this section by noting that an experimental TCP implementation known as "Vegas" was proposed in 1994 that implemented a congestion avoidance mechanism designed to avoid losses by reducing the window upon a detection of an increase in the RTT (which would indicate queues building along the path) [Brakmo et al., 1994]. TCP Vegas has not been shown to work well in a heterogeneous environment; in particular, in a satellite environment, it exhibits rather poor performance because it is very slow to probe for unused bandwidth [Zhang et al., 1997]. Because of the bias against longer RTT connections when congestion occurs, TCP implementations that are less aggressive (such as Vegas) will likely perform poorly when the satellite link is being used to access the wide-area Internet. Although some experiments suggest that more aggressive TCP connections can recoup some of this lost performance without adversely affecting other flows, there are significant deployment issues surrounding such mechanisms [Henderson et al., 1998].

## 5.7 Using multiple data connections

Before the advent of the window scale enhancements discussed above, one popular method of circumventing the throughput restrictions that small-windowed TCP connections provided was to write applications that launched multiple TCP connections in parallel and multiplexed data across them. This approach is sometimes colloquially known as "striping." Such applications typically see a substantial performance improvement, for both short transactions (since the application effectively receives an initial window size of N segments for N connections) and long transfers (because the aggregate window is large, and the impact of a single dropped segment reduces the throughput on only one of the parallel connections). An example application is described in [Allman et al., 1996]. While this technique demonstrates improved performance over satellite channels, the aggressive behavior has been shown to contribute to congestive collapse in shared networks [Floyd and Fall, 1999]. The IETF suggests that this solution may be safe and beneficial in private networks, but not in networks for which the end-to-end connections traverse the shared Internet [Allman et al., 2000].

## 5.8 Header compression

TCP/IP headers are typically 40 bytes long (IPv4, without options), and constitute a significant fraction of the overall bandwidth when packets are short (such as in voice-over-IP). Many of the fields in the headers do not change from packet to packet, a property that led to the definition of header compression techniques for TCP/IP that replace the headers with a smaller session identifier and allowed per-packet overhead to drop to between 3-5 bytes [Jacobson, 1990]. This compression can be extremely beneficial for bandwidth-limited point-to-point links. Extra space is saved when only the differences between slowly changing fields are sent, rather than the whole field. However, such "delta encoding" is much more sensitive to packet losses because the header decompression relies on the presence of previous packets. Algorithms that aim to resynchronize the compression endpoints in the event of a loss (such as the "twice" algorithm [Degermark et al., 1997]) are important in this case, especially since a rapid resynchronization is necessary to allow the fast retransmit/fast recovery algorithms to engage (and thereby avoid a coarse timeout).

While simulations show that header compression can improve throughput by 10-15% even over a wide range of bit error rates [Degermark et al., 1997], implementing header compression requires changes to the link layer protocol implementations at both the sending and receiving side of the link. Furthermore, the use of IP-level encryption will block the ability to compress the TCP headers.

## 5.9 TCP Performance Enhancement Proxy

Although TCP can work well over even GEO satellite links under certain conditions, we have illustrated that there are cases (e.g., Section 5.6) for which even the best end-to-end modifications cannot ensure good performance. Furthermore, in an actual network with a heterogeneous user population, users and servers cannot all be expected to be running satellite-optimized versions of TCP. This has led to the practice of splitting transport connections by inserting one or more "performance enhancing proxies" (IETF terminology) along the path. This concept is not new; satellite operators have deployed protocol conversion gateways for many years, and the IETF has produced an informational RFC summarizing various proxy techniques [Border et al., 2001]. The idea behind split connections is to shield high-latency or lossy network segments from the rest of the network, in a manner transparent to applications. TCP connections may be split in a number of ways. Figure 6.9 illustrates a common case, in which a gateway is inserted on the link between the satellite terminal equipment and the terrestrial network. On the user side, the gateway may be integrated with

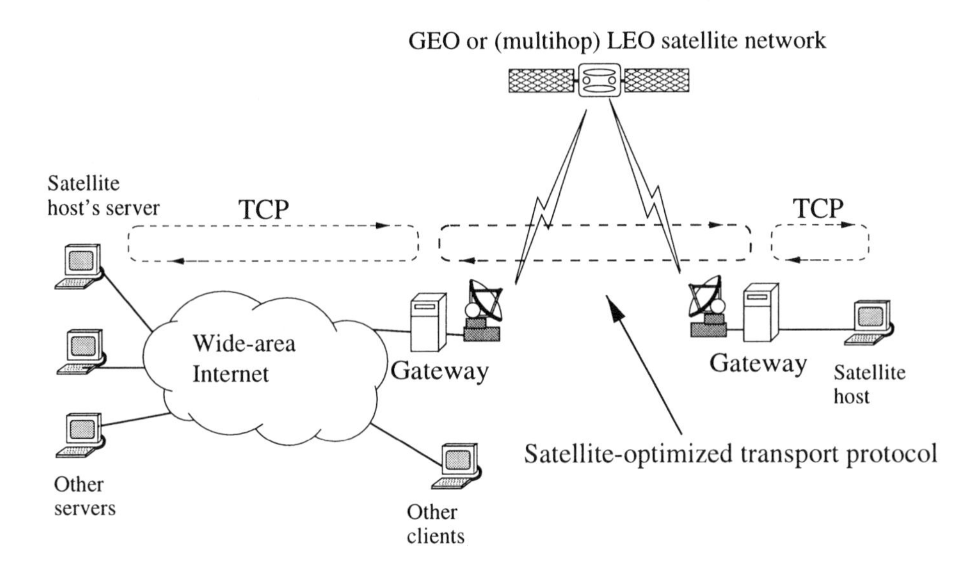


*Figure 6.9.* Future satellite networking topology in which a satellite-based host communicates with a server in the Internet through a satellite protocol gateway

the user terminal, or there may be no gateway at all. The goal is for end users to be unaware of the presence of an intermediate agent, other than improved performance. From the perspective of the host in the wide-area Internet, it is communicating with a well-connected host with a much shorter latency. Over the satellite link, a satellite-optimized transport protocol can be used. The next chapter provides more details on this subject, and recent publications provide experimental results (e.g. [Henderson and Katz, 1999], [Ishac and Allman, 2001]).

## 5.10 Additional protocols

We close this chapter by briefly introducing two related protocols that may outperform TCP over satellite channels. NASA's Jet Propulsion Laboratory has been leading the development of protocols optimized for reaching remote space vehicles. Transport connections to space probes could potentially operate with latencies on the order of minutes or hours. This effort has led to the development of a modified version of TCP known as the *Space Communications Protocol Standards– Transport Protocol (SCPS-TP)* for the general space environment [Durst et al., 1997]. SCPS-TP proposes a new TCP option which would enable several changes to basic TCP mechanisms, including the following: distinguishing between packet loss and packet errors (to react differently to the two events), using specially tuned TCP Vegas congestion avoidance algorithms, identifying link outage events, performing header compression, and using selective negative acknowledgments.

The Stream Control Transmission Protocol (SCTP) is an alternative transport protocol, originally designed for transport of signaling messages over the Internet but now contemplated for use with user data transfer, in place of TCP. SCTP offers a message-based data transfer service with the following characteristics: TCP-like congestion control, support for multi-homed hosts, more robustness against denial-of-service attacks, and support for optional out-of-order data delivery [Stewart et al., 2000]. SCTP's congestion control and loss recovery mechanisms closely resemble those of the TCP SACK with NewReno implementation described above, with the addition of congestion window growth due to byte counting, and hence the performance due to congestion control and loss recovery in a satellite environment should be similar.

## 6. SUMMARY

TCP has proven to be a very robust protocol in a variety of network environments. However, this chapter has illustrated that obtaining good performance using standard end-to-end TCP connections is very challenging in a GEO satellite environment. Thanks to enhancements such as TCP window scale, TCP performance over satellite channels is not necessarily relegated to the slow lane– the University of Kansas has been active in experimenting with specially-tuned TCP implementations over high-speed satellite channels available on the NASA Advanced Communications Technology Satellite (ACTS), which provides channels at up to OC-12 (622 Mb/s) rates [Charalambous et al., 1998]. However, challenges still remain in simultaneously achieving high throughput and bandwidth efficiency while exhibiting behavior that is considered "friendly" to other TCP connections sharing the end-to-end path. The next chapter in this book describes in more detail the performance benefits that can be obtained by using special performance-enhancing TCP proxies within satellite networks.

## References

Allman, M. (1999). TCP Byte Counting Refinements. *ACM Computer Communications Review*, 29(3):14–22.

Allman, M. et al. (2000). Ongoing TCP Research Related to Satellites. *Internet RFC 2760.*

Allman, M., Floyd, S., and Partridge, C. (1998). Increasing TCP's Initial Window. *Internet RFC 2414.*

Allman, M., Glover, D., and Sanchez, L. (1999). Enhancing TCP Over Satellite Channels using Standard Mechanisms. *Internet RFC 2488.*

Allman, M., Kruse, H., and Ostermann, S. (1996). An Application-Level Solution to TCP's Satellite Inefficiencies. *Proceedings of 1st Workshop on Satellite-Based Information Systems (WOSBIS '96).*

Balakrishnan, H. (1998). Challenges to Reliable Data Transport Protocols over Heterogeneous Wireless Networks. *Ph.D. Thesis, University of California, Berkeley.*

Balakrishnan, H., Padmanabhan, V., and Katz, R. (1997). The Effects of Asymmetry on TCP Performance. *Proceedings of Third ACM/IEEE MobiCom Conference*, pages 77–89.

Balakrishnan, H., Rahul, H., and Seshan, S. (1999). An Integrated Congestion Management Architecture for Internet Hosts. *Proceedings of ACM SIGCOMM Conference*, pages 175–87.

Border, J., Kojo, M., Griner, J., Montenegro, G., and Shelby, Z. (2001). Performance Enhancing Proxies Intended to Mitigate Link-Related Degradations. *Internet RFC 3135.*

Braden, R. (1994). T/TCP– TCP Extensions for Transactions, Functional Specification. *Internet RFC 1644.*

Brakmo, L., O'Malley, S., and Peterson, L. (1994). TCP Vegas: New Techniques for Congestion Avoidance. *Proceedings of ACM SIGCOMM '94*, pages 24–35.

Brakmo, L. and Peterson, L. (1995). Performance Problems in BSD4.4. TCP. *ACM Computer Communications Review*, 25(5):69–86.

Charalambous, C. et al. (1998). Experimental and Simulation Performance Results of TCP/IP over High-Speed ATM over ACTS. *Proceedings of Int. Conf. on Communications (ICC)*, 1:72–78.

Charalambous, C., Frost, V., and Evans, J. (1999). Performance of TCP Extensions on Noisy High BDP Networks. *IEEE Letters on Communications*, 3(10):294–299.

Degermark, M., Engan, M., Nordgren, B., and Pink, S. (1997). Low-Loss TCP/IP Header Compression for Wireless Networks. *ACM/Baltzer Journal on Wireless Networks*, 3(5):375–87.

Durst, R., Miller, G., and Travis, E. (1997). TCP Extensions for Space Communications. *Wireless Networks*, 3(5):389–403.

Fall, K. and Floyd, S. (1996). Simulation-based Comparisons of Tahoe, Reno, and SACK TCP. *ACM Computer Communications Review*, 26(3):5–21.

Floyd, S. (1991). Connections with Multiple Congested Gateways in Packet-Switched Networks, Part 1: One-way Traffic. *ACM Computer Communications Review*, 21(5):30–47.

Floyd, S. (1994). A Proposed Modification to TCP's Window Increase Algorithm. *Unpublished draft, cited for acknowledgement purposes only.*

Floyd, S. and Fall, K. (1999). Promoting the Use of End-to-End Congestion Control in the Internet. *IEEE/ACM Transactions on Networking*, 7(4):458–72.

Floyd, S. and Henderson, T. (1999). The NewReno Modification to TCP's Fast Recovery Algorithm. *Internet RFC 2582.*

Floyd, S. and Jacobson, V. (1992). On Traffic Phase Effects in Packet Switched Gateways. *Internetworking: Research and Experience*, 3(3):115–156.

Gavish, B. and Kalvenes, J. (1998). The Impact of Satellite Altitude on the Performance of LEOS-based Communication Systems. *Wireless Networks*, 4(2):199–212.

Hashem, E. (1989). Analysis of Random Drop for Gateway Congestion Control. *Report LCS TR-465, Laboratory for Computer Science, MIT, Cambridge, MA.*

Heidemann, J. (1997). Performance Interactions Between P-HTTP and TCP Implementations. *ACM Computer Communications Review*, 27(2):65–73.

Heidemann, J., Obraczka, K., and Touch, J. (1997). Modeling the Performance of HTTP Over Several Transport Protocols. *ACM/IEEE Transactions on Networking*, 5(5):616–630.

Henderson, T. (1999). TCP Performance over Satellite Channels. *Technical Report, University of California, Berkeley.*

Henderson, T. and Katz, R. (1999). Transport Protocols for Internet-Compatible Satellite Networks. *IEEE Journal on Selected Areas in Communications*, 17(2):345–59.

Henderson, T., Sahouria, E., McCanne, S., and Katz, R. (1998). On Improving the Fairness of TCP Congestion Avoidance. *Proceedings of IEEE Globecom '98 Conference.*

Hoe, J. (1996). Improving the Start-up Behavior of a Congestion Control Scheme for TCP. *Proceedings of ACM SIGCOMM '96 Conference*, pages 270–280.

Ishac, J. and Allman, M. (2001). On the Performance of TCP Spoofing in Satellite Networks. *Proceedings of IEEE MILCOM '01 Conference.*

Jacobs, I., Binder, R., and Hoversten, E. (1978). General Purpose Packet Satellite Networks. *Proceedings of the IEEE*, 66(11):1448–67.

Jacobson, V. (1988). Congestion Avoidance and Control. *Proceedings of ACM SIGCOMM '88 Conference*, pages 314–329.

Jacobson, V. (1990). Compressing TCP/IP Headers. *Internet RFC 1144.*

Jacobson, V., Braden, R., and Borman, D. (1992). TCP Extensions for High Performance. *Internet RFC 1323.*

Lakshman, T. and Madhow, U. (1997). The Performance of TCP/IP for Networks with High Bandwidth-Delay Products and Random Loss. *IEEE/ACM Transactions on Networking*, 5(3):336–350.

Lakshman, T., Madhow, U., and Suter, B. (1997). Window-based Error Recovery and Flow Control with a Slow Acknowledgment Channel: A Study of TCP/IP Performance. *Proceedings of INFOCOM '97*, pages 1199–1209.

Mathis, M. and Mahdavi, J. (1996). Forward Acknowledgment: Refining TCP Congestion Control. *Proceedings of ACM SIGCOMM '96 Conference*, pages 281–92.

Mathis, M., Mahdavi, J., Floyd, S., and Romanow, A. (1996). TCP Selective Acknowledgment Options. *Internet RFC 2018.*

Mogul, J. and Deering, S. (1990). Path MTU discovery. *Internet RFC 1191.*

Padmanabhan, V. and Katz, R. (1998). TCP Fast Start: A Technique for Speeding Up Web Transfers. *Proceedings of IEEE Globecom '98 Internet Mini-Conference.*

Padmanabhan, V. and Mogul, J. (1994). Improving HTTP Latency. *Proceedings of the Second International World Wide Web Workshop.*

Partridge, C. and Shepard, T. (1997). TCP Performance over Satellite Links. *IEEE Network*, 11(5):44–49.

Paxson, V. (1997). Automated Packet Trace Analysis of TCP Implementations. *Proceedings of ACM SIGCOMM '97 Conference*, pages 167–180.

Postel, J. (1981). Transmission Control Protocol. *Internet RFC 793.*

Samaraweera, N. (1999). Return Link Optimization for Internet Service Provision using DBS-S Networks. *ACM Computer Communications Review*, 29(3):4–13.

Seo, K. et al. (1988). Distributed Testing and Measurement across the Atlantic Packet Satellite Network (SATNET). *Proceedings of ACM Sigcomm '88*, pages 235–46.

Seshan, S., Stemm, M., and Katz, R. (1997). SPAND: Shared Passive Network Performance Discovery. *Proceedings of 1st Usenix Symposium on Internet Technologies and Systems (USITS '97).*

Stadler, J. (1997). A Link Layer Protocol for Efficient Transmission of TCP/IP via Satellite. *Proceedings of IEEE MILCOM '97*, pages 723–27.

Stevens, W. (1994). *TCP/IP Illustrated, Volume 1.* Addison Wesley.

Stevens, W. (1997). TCP Slow Start, Congestion Avoidance, Fast Retransmit, and Fast Recovery Algorithms. *Internet RFC 2001.*

Stewart, R. et al. (2000). Stream Control Transmission Protocol. *Internet RFC 2960.*

Suter, B., Lakshman, T., Stiliadis, D., and Choudhury, A. (1998). Design Considerations for Supporting TCP with Per-flow Queueing. *Proceedings of INFOCOM '98*, pages 299–306.

Touch, J. (1997). TCP Control Block Interdependence. *Internet RFC 2140.*

Zhang, Y., DeLucia, D., Ryu, B., and Dao, S. (1997). Satellite Communications in the Global Internet: Issues, Pitfalls, and Potential. *Proceedings of INET '97.*

Chapter 7

# TCP PERFORMANCE ENHANCEMENT PROXY

*The Industry's Best Practice*

Yongguang Zhang
*HRL Laboratories, LLC*

Abstract: It is commonly perceived that Internet applications, especially those based on TCP protocol, do not perform well over satellite networks. The root of this problem is the latency incurred in a satellite link. To stay competitive, the satellite industry has developed and deployed a practical solution, called TCP Performance-Enhancement Proxy (as known as "TCP Spoofing") to solve this problem. Although it has gained tremendous success in today's satellite networks, its use has always been controversial among the Internet community. Its critics blast it for breaking the end-to-end principle, an orthodox of the Internet. Lately, the increasingly pervasive IPsec applications have begun to threaten the performance of satellite networks by rendering TCP Performance-Enhancement Proxy useless for IPsec-protected traffic.

Key words: TCP, Performance-Enhancement Proxy, TCP Spoofing, IPsec.

## 1. INTRODUCTION

The Transmission Control Protocol (TCP) [Stevens, 1994] is used as *the* transport layer protocol by many Internet and intranet applications. It is vitally important to the success of a new networking technology that it sustains high performance for TCP applications. Satellite networks are no exception; they must maintain high TCP throughput to stay competitive and be accepted as a viable networking technology for the global Internet.

Today's satellite networks are implemented with geo-synchronous (GEO) satellites. They are typically a star topology with a large hub earth station at the center of the star and small terminals (satellite dishes) at the remote sites. The hub station is also an Internet gateway that connects the remote sites to the Internet backbone. There are many advantages to this technology compared with other terrestrial networks: high bandwidth, low infrastructure cost (no cable to lay), and available virtually everywhere. Therefore in early to mid 1990s when virtually the entire industry was entering the Internet market, many satellite companies (such as Hughes Electronics) have launched commercial Internet services with this technology.

Unfortunately, GEO satellite networks have much higher latency than terrestrial networks; it is in the order of 500ms versus 50ms. It is well understood that Internet applications, especially those based on TCP protocol, do not perform well under high latency environment. Many studies have shown that the throughput of a TCP session may be seriously limited if a satellite link is used (see the previous chapter). These are obviously bad news for the satellite industry, as they must produce a comparable TCP throughput performance to stay competitive with terrestrial networks.

Over the years, the industry has invested significantly in technologies that can battle such negative impact from latency and deliver comparable TCP performance. One such technology is called *TCP Performance-Enhancement Proxy* (or *TCP PEP* for short). In a nutshell, TCP PEP refers the technique that satellite gateways manipulate TCP traffic to influence the TCP's behavior. For example, a satellite gateway can "trick" the TCP sender into believing that the latency between sender and receiver were much smaller, so that it will send faster than it normally would. Since TCP PEP requires modifying the satellite gateways only and the changes can be made transparent to rest of the Internet, it has been used widely in many satellite network systems. It has gained tremendous success in improving the performance and usability of satellite networks.

Historically, TCP PEP is commonly referred as *TCP Spoofing* in the satellite industry. The commonly used technique in TCP Spoofing, which we will explain in detail later, is to have the satellite gateway pretend to be the intended destination and intercept all satellite-bound TCP connections for performance gains. Strictly speaking, TCP PEP is a broader term and TCP Spoofing is just one type of TCP PEP. Furthermore, the word "spoofing" is commonly associated with malicious intents such as hacking and intrusions in the networking industry. To "ratify" this activity, a neutral term "TCP PEP" is more appropriate and more acceptable. The intent and scope of TCP PEP is specified in the *PEP RFC* – a document that is formally approved and published by the Internet standardization body IETF as *RFC 3135* [Border et al., 2001].

In this chapter, we will present TCP PEP as the satellite industry's best practice to address the latency issue. We will analyze its pros and cons, and explain why, albeit its success in satellite networks, it is still controversial in the mainstream Internet community.

# 2. THE MOTIVATION

Among the many problems that TCP face in a high-latency environment such in the satellite networks (see Chapter 6), the TCP PEP work was mainly motivated by two of them: the TCP slow start problem and the historical small window size problem.

## 2.1 The Slow-Start Problem

TCP is known to have performance problem in long-delay networks. This is because TCP's closed-loop flow control depends on network feedback, which is embodied in ACK arrivals. The sooner the ACK arrives, the sooner the sender can be made aware of the network condition, and sooner it can transmit the next data packet. Therefore, longer delay implies slower response to network condition, and it will take TCP longer to reach the optimal throughput.

This phenomenon is manifested in a standard TCP mechanism called *slow-start* [Stevens, 1994]. When a TCP connection first starts up or after it stays idle for a long time, it needs to quickly determine the available bandwidth on the network. It does so by starting with an initial window size of one or two segments (usually 512 or 1024 byte segments), and then increasing the window size as acknowledgements arrive (meaning that packets are delivered successfully), until it reaches the network saturation state. On one hand, this "water-testing" procedure avoids overtaxing the network before it has a good assessment of the available bandwidth; on the other hand, TCP bandwidth utilization is sub-optimal during this period. Therefore, the shorter the TCP slow-start lasts, the better performance the TCP can achieve.

*Figure 7-1* illustrates the TCP slow-start procedure. The upper chart shows the time line of a long-lasting TCP session, starting from the time a connection is established. The Y-axis is the TCP sending rate and the shaped part is the amount of data this TCP session transmits over time. Initially, TCP sending rate starts from one segment at a time. Assuming that it incurs no packet loss (i.e., no congestion), the sending rate doubles after every round-trip (i.e., latency $\times$ 2). The start-slow procedure ends when the sending rate passes a pre-defined value called *ssthresh* (slow-start-threshold), which is expected at about one half of the available bandwidth. The double arrow in the chart marks this slow-start period. After the slow-start, the sending rate continues to increase but by one segment every round-trip only, until it reaches the available bandwidth. From the figure we can see that the TCP throughput is indeed sub-optimal in this period.

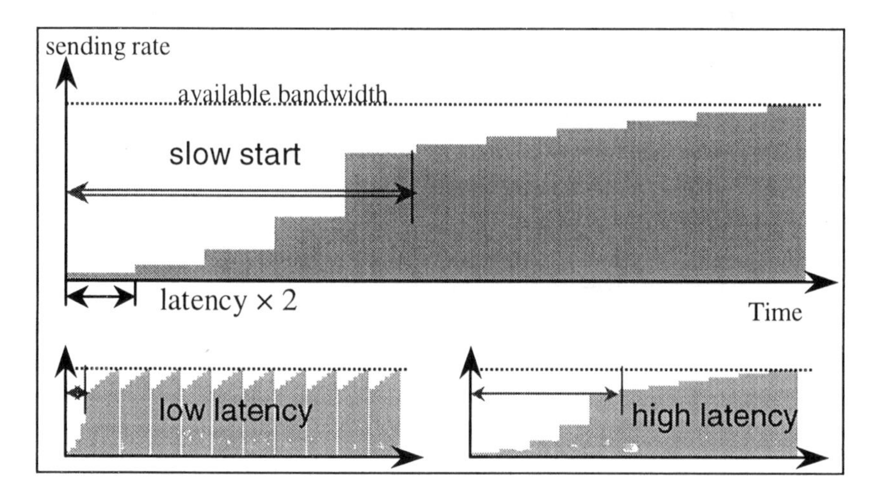


*Figure 7-1.* TCP slow-start illustrated

The length of this period can be roughly calculated as the following. Assuming the available bandwidth is $B$, the TCP segment size is $TSS$, and the round-trip time is $RTT$, then the total elapsed time for the initial slow-start period of a TCP connection will be

$$SlowStartTime = RTT \times (\log_2(B \times RTT/TSS)+1)$$

The average throughput of a TCP connection during this slow-start period will be

$$SlowStartThroughput = (2 \times B - TSS/RTT)/ \log_2(B \times RTT/TSS)$$

Although the TCP slow start usually does not last long in a terrestrial wired network with relatively small latency, the effect cannot be neglected when TCP is used over GEO satellite because the latency can be one order of magnitude higher (500ms vs. 50ms). The bottom two charts in *Figure 7-1* clearly illustrate this effect. If both networks have the same available bandwidth, and one has lower latency and the other higher latency, the TCP throughput is higher in the lower latency network. For satellite networks to compare favorably with terrestrial networks, the negative impact of slow-start must be resolved.

## 2.2 The Window Size Problem

The performance of TCP is also affected by the TCP window size. Technically, the TCP window size refers to the buffer in TCP sender for the data that has been transmitted but not yet acknowledged. That is, the TCP window size is the maximum amount of data that a TCP session can put into the network at any time. Therefore, the throughput of TCP is limited by the following factor, even if the available bandwidth can be higher:

*Throughput = Window Size / Round-Trip Time*

If the round-trip time is small, a smaller window size is less likely to be a problem, but for satellite networks, one needs to have a large window size to achieve optimal performance [Jacobson et al., 1992]. Furthermore, this equation assumes perfect flow condition. In practice, if there are losses, typically two to three times this window value is required to maintain high throughput.

For example, if we have a 1000ms round-trip time in a satellite network, and if we want to achieve 1Mbps throughput, we need to have a window size of at least 125KB under perfect condition, or more to deal with losses.

Historically, the window sizes in standard TCP and in many implementations were rather small. For example, when TCP PEP was in the design board in early 1990s, the TCP window sizes in most Internet clients were as small as 4KB [Stevens, 1994]. Even the Internet standard for TCP specified an upper limit of 64KB. To support a window size larger than 64 KB, an optional TCP Large Window Extensions (RFC 1323 [Jacobson et al., 1992]) would have to be implemented. Under this circumstance, it was obviously impractical for satellite network operators to require every user and Internet servers to implement and configure TCP Large Window. A transparent and "inside-the-network" solution was required.

Granted, today's Internet hosts generally have a larger window size and many operating systems have already implemented RFC 1323. However, some special configurations (like large socket buffers, described in Chapter 6) are still required to use them. It is therefore still desirable to have a transparent solution that does not require change to the server or the client.

# 3. THE PRACTICAL SOLUTION

There have been many researches on how to improve the performance of TCP over satellite networks [Allman et al., 1999; Allman et al., 2000]. However, many of them recommend modifications of the TCP protocol or implementation. Others require non-standard or uncommon configurations. These research results will ultimately help, if they can become a universally accepted operational standard. However, there is no guarantee that any of these can become an Internet standard eventually, and to become a standard it still has to go through many years of IETF process. Further, it is difficult if not impossible to change all Internet hosts with the new software and/or configurations in a short period of time. Therefore, the satellite network operators are forced to adopt a non-standardized but non-intrusive solution, which meets the following requirements:

- No change (software or configuration) is to be made at the server hosts in the Internet.
- No change is to be made at any Internet routers except the segment belonging to the satellite network.
- Changes can be made at users' computers, such as a new TCP stack, but they must be minimum and must be part of the software required to access the satellite networks. For example, if it requires new software/driver for accessing the satellite network, the new driver can conveniently change the TCP stack's behavior.
- In principle the changes should be transparent both to the Internet servers and to the users.

The solution is transparent TCP PEP, or what had been known as TCP Spoofing.

## 3.1 Basic Architecture

We now illustrate the transparent TCP PEP mechanism by contrasting the two graphs in *Figure 7-2*. Let's assume the TCP sender is the Server in the Internet and the TCP receiver is the User over the other end of the satellite link. If a standard TCP session spans over a satellite network, a TCP data packet will travels from the sender (Server) to the receiver (User), and a TCP ACK packet will travels back from the receiver to the sender. The perceived round-trip delay is long (upper graph in *Figure 7-2*).

The new architecture with TCP PEP deployment is illustrated at the lower graph in *Figure 7-2*. A *PEP agent* is installed at the satellite uplink gateway between the Internet and the satellite network. It intercepts every TCP packets between these two networks and analyzes their headers for flow information such as sequence numbers and acknowledged segments. When a data segment passes through from the sender to the receiver, the PEP agent will generate a "premature" acknowledgement to be returned to the sender, without waiting for the data segment to travel to the intended receiver or for the receiver to return a "real" acknowledgement. The premature acknowledgements are specially formatted to be indistinguishable from the real acknowledgements – except that they arrive at the sender sooner. The perceived round-trip delay is therefore shortened significantly. This gives the sender the illusion of a low-latency network and the TCP slow-start phase can progress much quicker.

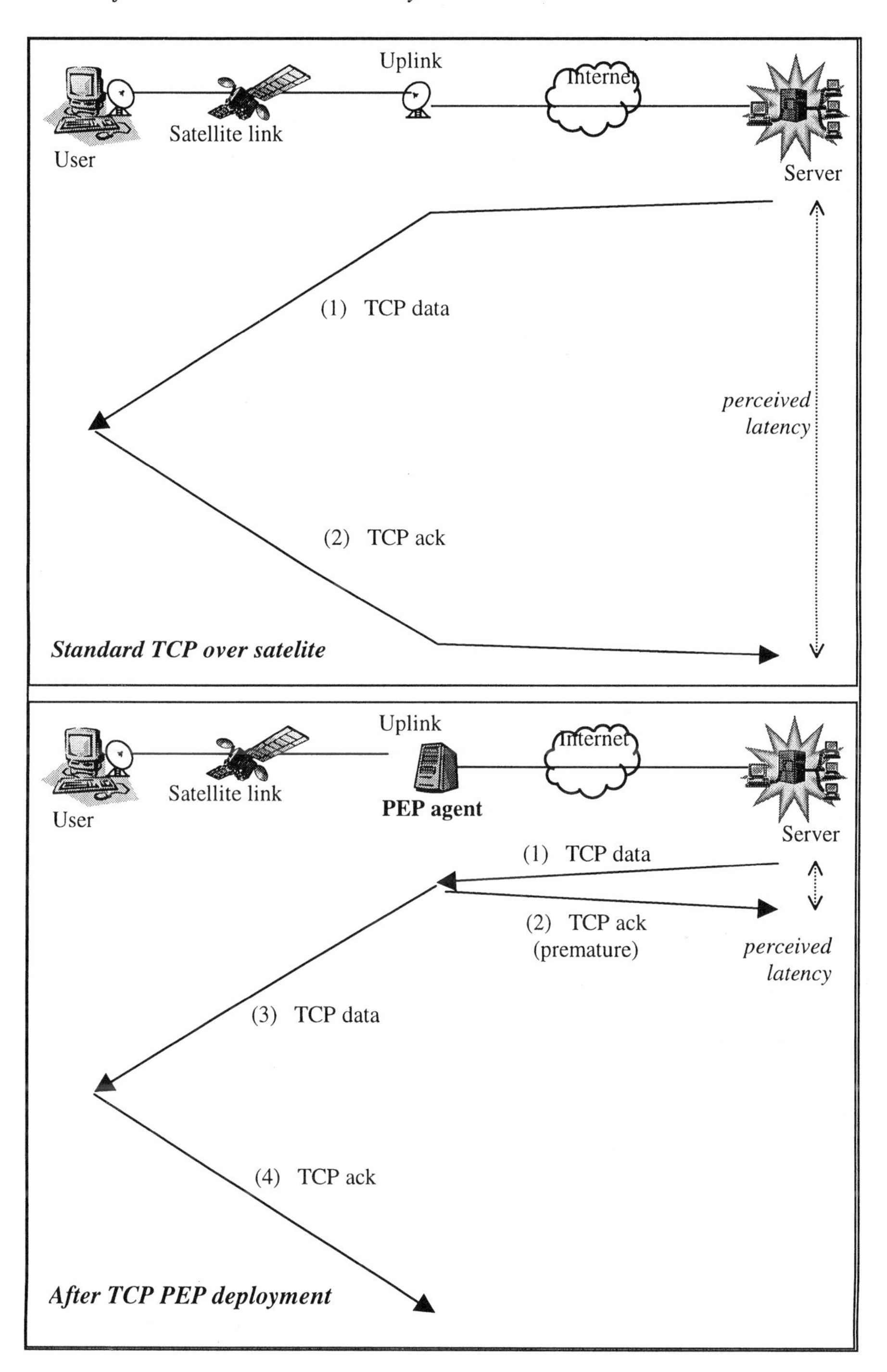


*Figure 7-2.* Effect of TCP PEP

The PEP agent buffers and tracks all the TCP segments in transit (i.e., has arrived at the gateway but not acknowledged by the receiver). When the real acknowledgements arrive, the PEP agent suppresses them to prevent duplication. If an acknowledgement never arrives before the normal TCP timeout (for example due to downstream congestion), the PEP agent retransmits the lost segment from its buffer. The PEP agent can also act on other indications of packet loss, such as "three duplicated ACKs". In practice, the PEP agent will employ sophisticated scheduler and coordinate with real-time resource management in the satellite network to avoid congestions. It will reduce or suspend the premature acknowledgements if a queue is building up at the satellite uplink.

The throughput improvement that TCP PEP can achieve depends on many factors [Ishac and Allman, 2001]. They include the latency of the satellite link, the delay between the satellite gateway and the Internet host, the satellite channel bandwidth, the congestion state in the Internet, and the amount of buffer space at the PEP agent as well. There are several performance studies on TCP PEP [Arora et al., 1996; Henderson and Katz, 1999; Bharadwaj 1999; Ishac and Allman, 2001; Ehsan et al., 2002]. The next chapter of this book will present one such study.

It should be noted that the concept of "spoofing" TCP packets is not new; similar ideas have been employed in different contexts and applications. Satellite network is perhaps the most elaborated application.

## 3.2 Example: Deployment in HNS DirecPC

Hughes Network System DirecPC™ network is one of the pioneers to deploy TCP PEP for improved performance. Although the network architecture has evolved since the service inception at 1994 (e.g., it now supports two-way satellite links in addition to modem return), the role of TCP PEP remains critical to the operation and success of this service. Even TCP PEP has improved over time, from the initial study [Falk, 1994] to RFC-compliant [Border et al., 2001], but the principle remains the same. Here, we will study the initial deployment of TCP PEP in the DirecPC™ Turbo Internet™ product. The rationality for its deployment is still true to the later versions of DirecPC, or to other satellite networks.

We first explain how DirecPC works. The DirecPC network has three major parts: the satellite link (the satellite and the allocated channel), a DirecPC gateway, and many DirecPC terminals. The DirecPC gateway is located near the satellite uplink and it serves as the gateway between the DirecPC network and the Internet. A DirecPC terminal is actually the user's personal computer equipped with a special satellite antenna interface and special DirecPC software. A DirecPC terminal has two network interfaces:

one attached to a receive-only satellite dish (via PCI or USB interface), and the other being a modem. The DirecPC terminal software uses a special device driver to combine the two interfaces and make them appear as one virtual device. It sends all outgoing traffic (IP packets) to the DirecPC gateway through the modem connection, and receives incoming IP packets from the receive-only satellite interface. The DirecPC gateway is responsible for forwarding traffic from DirecPC terminals to the Internet, and for transmitting traffic to DirecPC terminals via the satellite link. In the commercial offering, DirecPC uses one or more Ku-band transponders for this service (each transponder offering 12-24Mbps bandwidth). To be competitive with the terrestrial services (such as ISDN or cable modem), DirecPC determines that it must be able to support at least 400Kbps for each TCP flow.

Without TCP PEP, DirecPC would suffer the same problems as we listed earlier. For example, study has shown that under standard configuration the maximum throughput that a TCP could achieve over the DirecPC network was below 120Kbps [Arora et al., 1996]. This is well below the 400Kbps requirement and far below the 12Mbps potential.

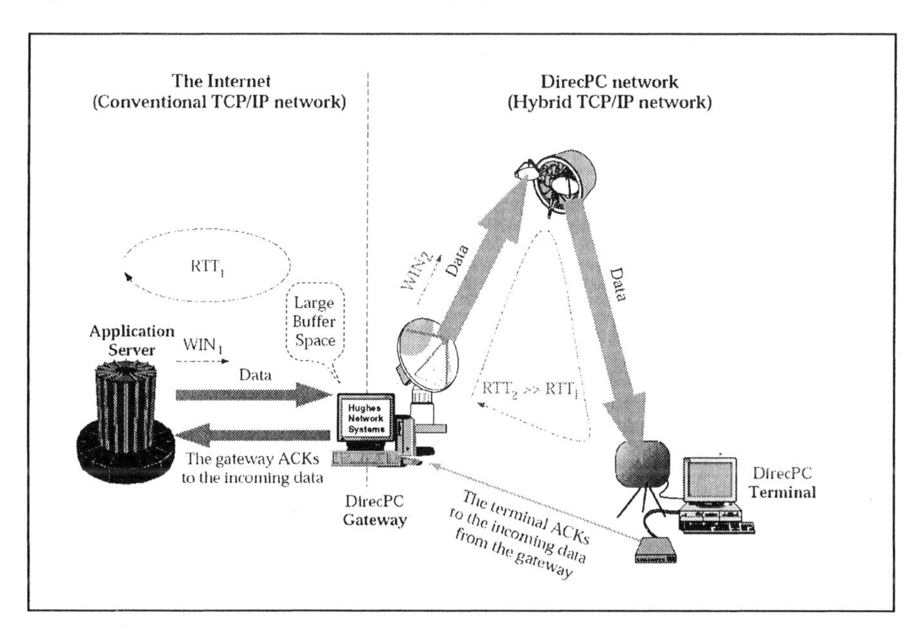


*Figure 7-3.* HNS DirecPC TCP PEP architecture (source: [Arora et al., 1996])

*Figure 7-3* illustrates the conceptual understanding of TCP PEP deployed in DirecPC. Here, the TCP sender is at the Application Server at left hand side and the TCP receiver is at the DirecPC Terminal at the right hand side. The normal TCP loop is from the Application Server to DirecPC

Terminal via DirecPC Gateway and the satellite link, and return via the modem link and the DirecPC gateway. The round-trip time is $RTT_1+RTT_2$ where $RTT_2>>RTT_1$.

With TCP PEP, the DirecPC Gateway terminates this loop with premature ACKs to the Application Server as soon as the TCP data packets arrive at the Gateway. To the TCP sender at the Application Server, this effectively reduces the round-trip time to $RTT_1$ and hence shortens the slow-start period and increases the throughput, even if the sender continues using the small window size.

At the same time, the DirecPC Gateway starts a separate TCP loop with the TCP receiver at the DirecPC Terminal. Although the latency is still high ($RTT_2$), but it uses a special traffic control mechanism to keep the throughput optimal. This special flow control is highly coupled with the satellite uplink scheduling, so that at anytime the DirecPC gateway knows the exact bandwidth allocated for each flow. This knowledge deems TCP slow-start unnecessary. Combining this with a much larger window size, the second TCP loop also archives high throughput.

Since the DirecPC Gateway has to generate ACKs for each data packet from the first loop and keep the packet until it being ACKed by the second loop, it requires large buffer space to store these data packets. Therefore, it is critical for the DirecPC Gateway to have large memory and a good memory management system. To ensure scalability, a server farm approach is used. At the DirecPC Gateway, the PEP agent function is distributed among a farm of server machines, each serving a subset of the DirecPC terminals. The number of terminals that a server can serve can be determined through analytical and experimental studies. This way, the number of users can scale by adding server machines to the farm.

With TCP PEP, the performance of TCP over DirecPC network has been improved significantly. DirecPC has since been able to support 400Kbps for a TCP flow [Arora et al., 1996].

## 3.3 Alternative Architecture and Mechanisms

While the mechanism explained in this section is still the predominant form of TCP PEP in satellite networks, other mechanisms can be used as well [Border et al., 2001]. For example, the following techniques are very common in TCP PEP:

- ACK Spacing. In environments where ACKs tend to bunch together, ACK spacing can be used to smooth out the flow of TCP acknowledgments traversing a link. This eliminates bursts of TCP data segments and hence reduces the buffer requirement at the satellite gateway (to hold these TCP packets).

- ACK Filtering. On paths with highly asymmetric bandwidth (such as the DirecPC network), the TCP ACKs flowing in the low-speed return path may get congested if the asymmetry ratio is high enough. ACK filtering can use TCP's "accumulated ACK" feature to delete ACKs on one side of the link and if necessary reconstruct the deleted ACKs on the other side of the link [Balakrishnan et al., 1999]. Actually, this mechanism has been implemented in the DirecPC's TCP PEP [Arora et al., 1996].

Depending on the configuration and topology of the satellite networks, there can be different architecture for designing TCP PEP. For example, if the satellite link is a transit link between two networks, a three-way TCP PEP architecture can be used (*Figure 7-4*).

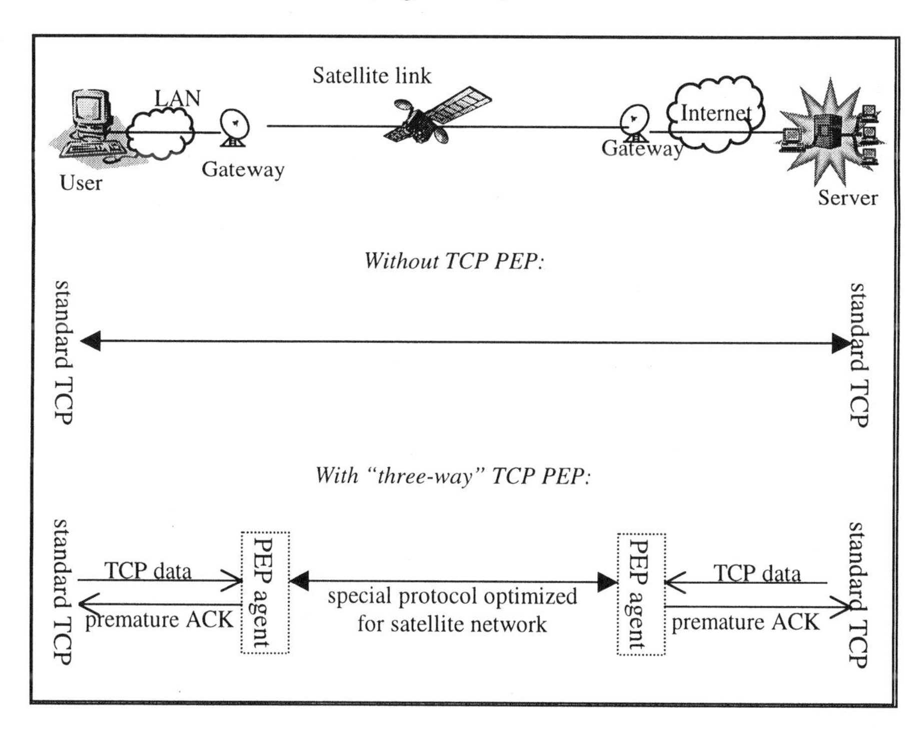


*Figure 7-4.* Three-way TCP PEP architecture

Besides transparent TCP PEP, non-transparent TCP PEP has the same architecture and same benefits. However, it requires explicit configurations on the users or servers (hence the non-transparence).

There are several companies selling commercial products based on TCP PEP. One such product is SkyX Gateway from Mentat. Although detail architecture is different in different systems, the basic principle remains same or similar to what has been described in this section.

# 4. THE BIG ARGUMENT

TCP PEP has brought tremendous competitive advantages to satellite networks. However, it has also been the target of strong criticisms ever since its inception. The argument is that TCP PEP breaks "the end-to-end principle" [Saltzer et al., 1984].

The end-to-end principle can be briefly described as the following: as a first principle, certain required end-to-end functions can only be correctly performed by the end systems themselves. Reliable message delivery is an end-to-end responsibility, and TCP, as a reliable transport protocol, should have the end-to-end semantics: when an ACK packet comes back to the sender, it is a sure indication that the corresponding data segments have arrived at to the receiver successfully and without error. When using TCP as the transport protocol, distributed systems may rely on this semantics for its proper function.

TCP PEP indeed breaks this semantics. When the satellite gateway generates a premature ACK packet, the corresponding data packet is only in transit. By the time this ACK packet arrives at the sender, the data packet may or may not reach the receiver yet, or worse, the data packet may be lost in transmit, or corrupted with transmission errors. But once the sender receives this ACK packet, it will consider the delivery successful, even though this may or may not have happened, and there is no guarantee it will happen. This is the main reason why the Internet community has been strongly against the use of this type of TCP PEP and why similar techniques have not been adopted outside of the satellite network industry.

The general issues of TCP PEP or other PEPs breaking the end-to-end principle are best analyzed and explained in the PEP RFC [Border et al., 2001]. In this section, we look at two issues, end-to-end reliability and fate sharing, in the context of TCP PEP. In the next section, we will focus on the end-to-end security and the IPsec issue.

## 4.1 The End-to-end Reliability Issue

TCP PEP has included mechanisms to remedy, to some degree, the negative impact of losing the end-to-end semantics. For example, TCP PEP implements a reliability mechanism between the PEP agent and the receiver: the satellite gateway will keep a buffer of all the packets that have not been acknowledged by the receiver, and it will retransmit those packets that have been lost or corrupted. However, this cannot replace the end-to-end semantics. First, the correct delivery of a TCP data packet is now an after-fact. The ACK that PEP agent generates now becomes a "promise" rather than a "certification" of delivery. Second, although the PEP agent can keep retransmitting a lost packet, there are situations that it cannot remedy, such as the receiver crashing, or even the PEP agent itself crashing. Therefore, it is impossible for TCP PEP to guarantee end-to-end reliability from inside the network.

Does this mean that TCP PEP is fundamentally flawed and should not be used? The question is whether the application is (or *should* be) prepared to deal with the loss of reliability guarantee. The end-to-end principle suggests that lower levels are not the right place to provide higher-level functions (such as reliability) [Saltzer et al., 1984]. Border et al. further argue that the application must be prepared to deal with the reliability issue itself, because it is a violation of the end-to-end principle at the application-level to depend solely on the lower level (TCP) for end-to-end reliability function [Border et al., 2001]. From this point of view, the applications should at least implement some form of checks and remedies for the rare cases of TCP failures. In fact, many Internet applications do include such functions, such as Internet Mail Transfer Agents (like sendmail, which retries a TCP connection if it fails). Others may rely on user behaviors, such as refreshing a web page that fails to load the first time. With this type of reliability in the application level, TCP PEP failure is remediable. This is why many Internet applications have been running properly in satellite networks with TCP PEP, and after several years of deployment, services like DirecPC has broken few applications. Therefore, we can conclude that TCP PEP is not flawed; the loss of end-to-end semantics in TCP PEP does not effect the applications that have higher-level end-to-end semantics.

## 4.2 The Fate Sharing Issue

Fate sharing refers to the design principle that state information describing an end-to-end connection (such as TCP) should be maintained only at the end systems and not inside the network [Clark, 1988]. This way, the loss of such information is only acceptable if an end system itself is lost. In the case of TCP, it also suggests that a TCP connection will be broken only if there is no path between the two ends. TCP PEP violates this rule too. The PEP agent maintains significant state for each TCP connection. If the PEP agent crashes, all the states for all connections are lost. The connections will have to terminate even if there are alternative paths.

However, TCP PEP is often deployed in a network configuration that there is no alternative path, such as the DirecPC case. If the satellite gateway crashes, all connections will terminate with no exception, whether the satellite gateway has a PEP agent or not. On the one hand, the PEP agent should be made as reliability as possible. In fact, satellite network operators have invested significant in reliability technique to keep TCP PEP state persistent, to survive failures and to recover rapidly. On the other hand, users of satellite networks should be aware of this implication. They should be prepared to accept the risk of losing TCP connections for the benefit of TCP PEP. Giving that TCP connections can be lost for other reasons and most applications have application-level reliability mechanisms (as we have described earlier), the impact is generally low.

## 5. THE "SHOW STOPPER"?

Another negative impact of losing end-to-end semantics in TCP PEP is its conflict with IPsec [Kent and Atkinson, 1998a] – a standard for secure communication over the Internet. It is also widely perceived that users will have to give up security in satellite networks, or, if the users choose to run IPsec or VPN (Virtual Private Network), they will suffer serious performance degradation in satellite networks.

So, is IPsec the "show-stopper" for TCP PEP, or for that matter, the current satellite networks? This section will discuss this issue.

### 5.1 Conflicts between IPsec and TCP PEP

We start with a brief introduction to IPsec [Kent and Atkinson, 1998a]. IPsec addresses the security issues at the IP layer. The fundamental concept is as followed. The path of an IP datagram between the source and destination is divided into three segments (see *Figure 7-5*) – the protected and trusted local network at the source (e.g., a company's private LAN), the untrustworthy public Internet segment, and the protected and trusted local network at the destination. IPsec places a security gateway (G1 and G2) at each boundary of trusted and untrustworthy networks. Initially, G1 establishes a security association with G2, meaning a security relationship that involves negotiation of security services and shared secrets, etc. Before an IP datagram (from S to D) is sent to the untrustworthy Internet, the security gateway (G1) encrypts and/or signs the datagram using an IPsec protocol. When it reaches the security gateway at the destination side (G2), the datagram is decrypted and/or verified, before it is forwarded for the destination (D). Optionally, the trusted local network in one or both side can be omitted, and the source or destination host can perform encryption, authentication and other security-gateway functions itself.

*Figure 7-5.* IPsec system model

IPsec uses two protocols to provide traffic security – AH (Authentication Header) [Kent and Atkinson, 1998b] and ESP (Encapsulating Security Payload) [Kent and Atkinson, 1998c]. AH provides integrity and authentication without confidentiality; ESP provides confidentiality, with optional integrity and authentication. Each protocol supports two modes of use: transport mode and tunnel mode. Transport mode provides protection primarily for upper layer protocols, while in tunnel mode the protection applies to the entire IP datagram.

The granularity of security protection in IPsec is at the datagram level. IPsec treats everything in an IP datagram after the IP header as one integrity unit. Usually, an IP datagram has three consecutive parts – the IP header (for routing purpose only), and the upper layer protocol headers (e.g., the TCP header), and the user data (e.g., TCP data). In transport mode, an IPsec header (AH or ESP) is inserted in after the IP header and before the upper layer protocol header to protect the upper layer protocols and user data. In tunnel mode, the entire IP datagram is encapsulated in a new IPsec packet (a new IP header followed by an AH or ESP header). In either mode, the upper layer protocol headers and data in an IP datagram are protected as one indivisible unit (see *Figure 7-6*).

TCP PEP operates on two pieces of state information stored in the headers of a TCP packet. They are TCP flow identification and sequence numbers within the flow. TCP flow identification is used to segregate TCP sessions for each TCP packet. It consists of source and destination IP addresses (both stored in IP header), as well as source and destination port numbers (both stored in TCP header). The sequence numbers are used to match acknowledgements with the data segments. This number is stored in TCP header. Without these two pieces of information, TCP PEP will not function.

When a TCP session is transported by an IPsec ESP mode, the TCP header is encrypted inside the ESP header. It is thus impossible for an intermediate gateway outside sender or receiver's security enclaves to analyze an IPsec header to extract TCP flow identification and sequence number. This will, inevitably, break the TCP PEP mechanisms. The PEP agent cannot obtain the information needed to generate acknowledgements or to retransmit data segments.

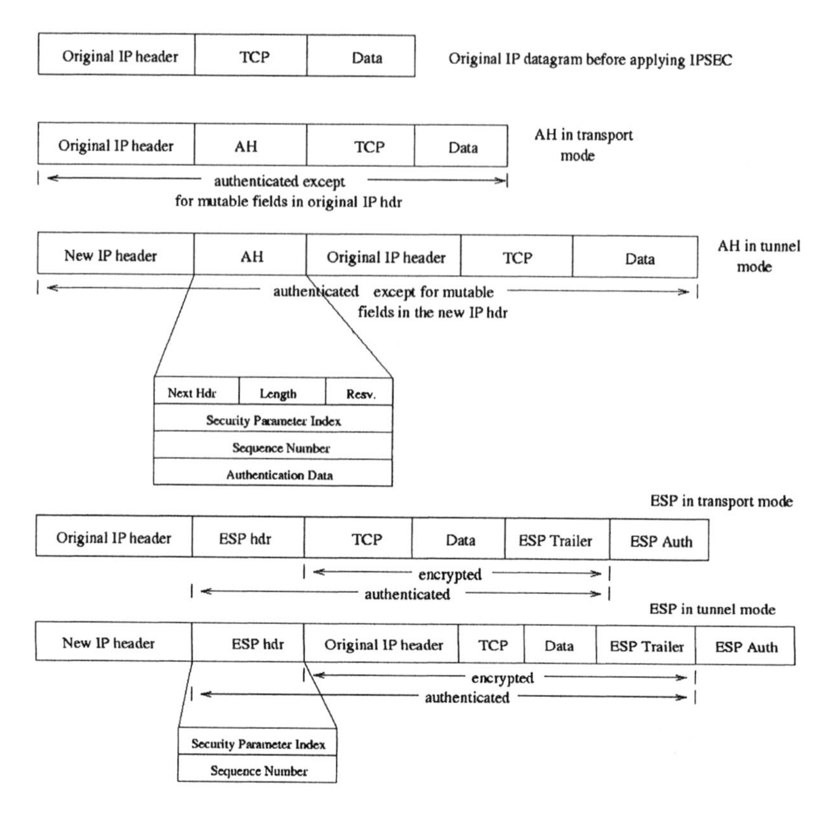


*Figure 7-6.* Protocol format for IPsec-protected IPv4 TCP packet

In a satellite networks, if IPsec is deployed end-to-end between the Internet server and the user host and if the TCP sessions are protected in this IPsec, the TCP PEP mechanism will not be able to provide performance enhancement. These TCP sessions will suffer the same problems described in Section 2.

## 5.2 The End-to-end Security Issue

TCP PEP critics have argued that the conflicts between TCP PEP and IPsec are precisely the reason why TCP PEP is a bad idea. However, the issue is more complicated than a simple mutual exclusion between TCP PEP and end-to-end security.

First, users of satellite network can choose to implement end-to-end security at a higher layer than IP, such as using the SSL library for transport-layer security (SSL or TLS [Rescorla, 2001]). These higher-layer security protocols are alternatives to IPsec. Further, they operate above TCP and work well with TCP PEP. In fact, many Internet applications already implement security with SSL or TLS, such as most web browsers (using HTTPS protocol), mail programs (using SIMAP and SPOP), etc.

Another issue is to define the realm of trust. Consider a typical satellite network in *Figure 7-7*, there are three entities: the User (end host), the Gateway, and the Server (end host), and two segments in the communication path: the satellite network between the User and the Gateway, and the public Internet between Gateway and the Server. The satellite network segment is operated and managed by a satellite network operator. If the users can trust this entity to provide proper security, it can put the PEP agent (at the Gateway) within its realm of trust. For example, the user can use the link-layer security mechanisms (such as satellite channel encryption) to send datagram over the satellite link, and use IPsec between the Gateway and the Server. Optionally, it can use another IPsec between User and the Gateway for additional protection over the satellite link.

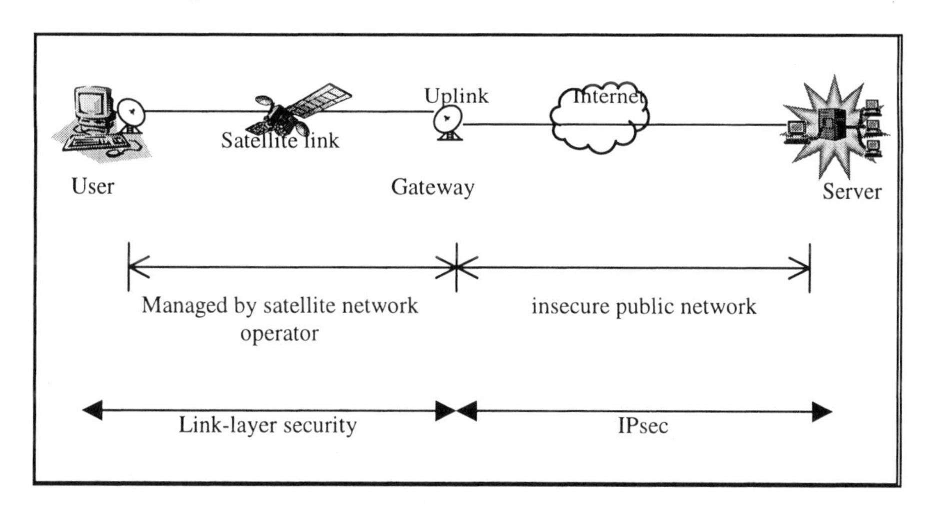


*Figure 7-7.* The realm of trust in a satellite network

In any case, the PEP agent must be put in the realm of trust in order for TCP PEP to work with IPsec. With the current IPsec standard, this requires the satellite Gateway to have access to TCP header, which is insider the IP payload, together with the TCP data (user data). That is, it places complete trust at the Gateway host, which may or may not be desirable.

To summarize, TCP PEP can be used together with end-to-end security. The following is a list of possible configurations:

- Use application-level security instead of IPsec, such as SSL or TLS [Rescorla, 2001].
- Use link-level security protection provided by the satellite network and use IPsec between satellite Gateway and the Internet host.

- Use different security configurations for different applications, e.g., use end-to-end IPsec for UDP or short-transaction type TCP flows (which may not need TCP PEP), and longer TCP flows with SSL.

In any case, the end users should be informed about this issue so that they can make an educated choice about end-to-end security and TCP PEP.

## 5.3 Researches on Resolving the Conflicts with IPsec

Although we recommend higher-layer security as it is not conflicting with TCP PEP, it is undeniable that implementing security in IP layer has its benefits. For example, by rendering the security services in a transparent manner, IPsec can relieve software developers from the need of implementing same security mechanisms at different layers or for different Internet applications. Further, IPsec is arguably the best mechanism for Virtual Private Networks (VPN) and secure remote accesses. As a result, IPsec deployment will become more and more popular.

There have been several research efforts to provide a solution that can work with both TCP PEP and IPsec (such as [Bellovin, 1999; Zhang and Singh, 2000]). One such approach is to develop a multi-layer security protection scheme [Zhang and Singh, 2000] for IPsec. The idea is to divide the IP datagram into several parts and apply different protection to different part. For example, the TCP payload is encrypted between two end points while the TCP/IP header is encrypted but accessible to two end points plus certain routers in the network. It is therefore possible to grant trusted intermediate routers a secure, controlled, and limited access to a selected portion of IP datagram, while preserving the end-to-end security protection to user data.

When we use this scheme with TCP PEP, we divide the IP datagram payload into two zones: TCP header and TCP data (see *Figure 7-8*). TCP data uses an end-to-end protection with keys shared only between sender and receiver. TCP header uses a separate protection scheme with keys shared among sender, receiver, and the satellite gateway (a trusted intermediate node). This way, no node in the network can access the TCP data, and no node except the satellite gateway can access the TCP header. The satellite gateway still has access to the TCP header to perform enhancement: to generate premature but acceptable TCP ACK packets.

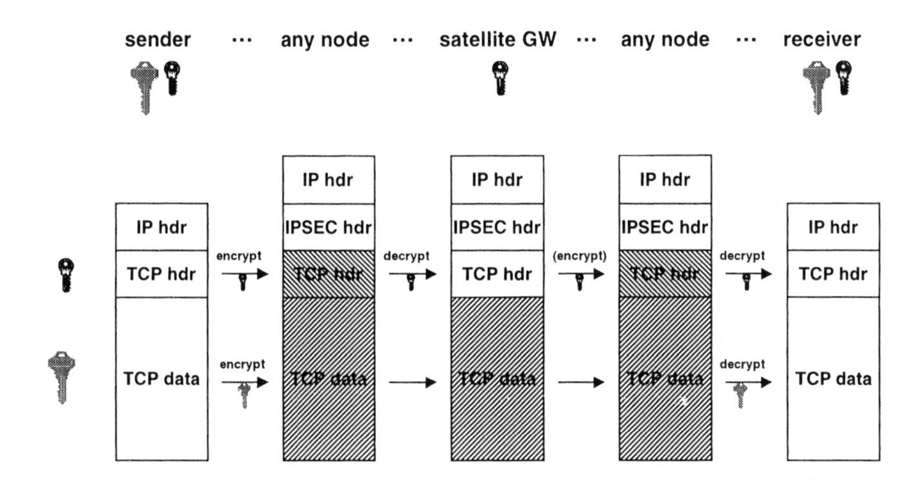


*Figure 7-8.* Multi-layer protection model for TCP (source: [Zhang and Singh, 2000])

# 6. CONCLUSION

Although being controversial, TCP PEP has provided a practical solution to the performance problem in the use of TCP over satellite networks. It has been widely used in today's commercial satellite networks and has become indispensable to their satisfactory performance.

TCP PEP does have some undesirable implications when it comes to the end-to-end argument. There is however little choice for improvement; the high latency in a satellite network is not something that can be easily compensated in link layer design. To put an end to the end-to-end argument around TCP PEP, the PEP RFC [Border et al., 2001] has the following compelling argument:

We believe that the end-to-end principle in designing Internet protocols should be retained as the prevailing approach and PEPs should be used only in specific environments and circumstances where end-to-end mechanisms providing similar performance enhancements are not available. In any environment where one might consider employing a PEP for improved performance, an end user (or, in some cases, the responsible network administrator) should be aware of the PEP and the choice of employing PEP functionality should be under the control of the end user, especially if employing the PEP would interfere with end-to-end usage of IP layer security mechanisms or otherwise have undesirable implications in some circumstances. This would allow the user to choose end-to-end IP at all times but, of course, without the performance enhancements that employing the PEP may yield.

## REFERENCE

Allman, M., Glover, D. and Sanchez, L. (1999). Enhancing TCP over satellite channels using standard mechanisms, IETF RFC 2488.

Allman, M., Dawkins, S., Glover, D., Griner, J., Henderson, T., Heidemann, J., Kruse, H., Ostermann, S., Scott, K., Semke, J., Touch, J. and Tran, D. (2000). Ongoing TCP research related to satellites, IETF RFC 2760.

Arora, V., Suphasindhu, N., Baras, J. and Dillon, D. (1996). *Effective Extensions of Internet in Hybrid Satellite-Terrestrial Networks*, University of Maryland Technical Report CSHCN 96-2.

Balakrishnan, H., Padmanabhan, V., and Katz, R. (1999). The effects of asymmetry on TCP performance, *ACM Mobile Networks and Applications (MONET) Journal,* 4(3).

Bellovin, S. (1999). *Transport-friendly ESP (or layer violation for fun and profit),* IETF-44 Meeting, TFESP-BOF.

Bharadwaj, V. (1999). *Improving TCP Performance over High-Bandwidth Geostationary Satellite Links*, University of Maryland Technical Report ISR-MS-99-12.

Border, J., Kojo, M., Griner, J., Montenegro, G., and Shelby, Z. (2001). Performance enhancing proxies intended to mitigate link-related degradations, IETF RFC 3135.

Clark, D.D. (1988). The Design Philosophy of the DARPA Internet Protocols, *Proceedings of ACM SIGCOMM'88.*

Ehsan, N., Liu, M., and Ragland, R. (2002). Measurement based performance analysis of Internet over satellite, In *Proceedings of 2002 International Symposium on Performance Evaluation of Computer and Telecommunication Systems.*

Falk, A. D. (1994). *System design for a hybrid network data communications terminal using asymmetric TCP/IP to support internet applications.* M.S. Thesis (CSHCN MS 94-2), University of Maryland.

Falk, A., Arora, V., Suphasindhu, N., Dillon D., and Baras, J.S. (1995). Hybrid Internet access, in *Proceedings of Conference on NASA Centers for Commercial Development of Space*, also: University of Maryland Technical Report CSHCN 95-7.

Henderson, T. and Katz, R. (1999). Transport Protocols for Internet-Compatible Satellite Networks. *IEEE Journal on Selected Areas in Communications*, 17(2):345-59.

Ishac, J. and Allman, M. (2001). The performance of TCP Spoofing in satellite networks, In *Proceedings of IEEE MILCOM 2001.*

Jacobson, V., Braden, B., and Borman, D. (1992). TCP extensions for high performance, IETF RFC 1323.

Kent, S., and Atkinson, R. (1998a). Security architecture for the Internet Protocol, IETF RFC 2401.

Kent, S., and Atkinson, R. (1998b). IP Authentication Header, IETF RFC 2402.

Kent, S., and Atkinson, R. (1998c). IP Encapsulating Security Payload (ESP), IETF RFC 2406

Rescorla, E. (2001). *SSL and TLS: Designing and Building Secure Systems*, Addison-Wesley.

Saltzer, J.H., Reed, D.P., and Clark, D.D. (1984). End-To-End Arguments in System Design, *ACM Transactions on Computer Systems*, 2(4):277-288.

Stevens, W.R. (1994). *TCP/IP Illustrated Volume I - The Protocols*, Addison-Wesley, Reading, Massachusetts.

Zhang, Y. and Singh, B. (2000). A multi-layer IPsec protocol, *Proceedings of the 9th USENIX Security Symposium*.

Chapter 8

# PERFORMANCE EVALUATION OF TCP SPLITTING OVER SATELLITE

Mingyan Liu
*Electrical Engineering and Computer Science, University of Michigan*

**Abstract** In this chapter we evaluate the performance of using TCP spoofing/splitting over satellite via a model-based study and a measurement-based study. Through analysis based on a few mathematical models we attempt to develop a basic understanding of the properties of TCP dynamics when connection spoofing is used, and further identify conditions under which using spoofing provides significant or marginal performance gain. Our measurement results are obtained from a commercial direct broadcast satellite system that uses TCP splitting at the network operating center (NOC). In particular these results show the performance of TCP splitting in combination with web caching. In both the model-based and the measurement-based studies we explore the effect of various factors including file size, congestion, and connection asymmetry. We use results and analysis from our models to better explain observations from the measurement. Finally we discuss the implications our findings and conclusions have on the design, deployment and provisioning of systems using TCP spoofing/splitting.

**Keywords:** Proxy, TCP spoofing/splitting, performance evaluation, caching, connection asymmetry

## 1. INTRODUCTION

In this chapter we examine the performance of using TCP connection splitting/spoofing [1] over satellite under various scenarios using both a model-based approach and a measurement-based approach. The goal is three-fold: to obtain a good understanding of the conditions under which using TCP splitting provides performance gain; to investigate the use of TCP splitting as a general solution to problems involving heterogeneous links and large amount of traffic; and more importantly, to apply such understanding to system-level designs. While the measurement-based study has a more realistic setting, it is conducted with a specific example system over which we have limited control. On the other hand, the model-based analysis is parameterized and therefore is more generally applicable, e.g., to scenarios where all connections are terrestrial and do not involve a satellite channel. The analysis aims to illustrate the underlying mechanism of TCP spoofing/splitting and the change in TCP dynamics in the presence of a splitting proxy. Thus the analytical results also help us interpret and understand the observations from measurement data.

The performance of TCP over heterogeneous medium such as those including satellite and wireless links has been extensively studied over the past few years. Proposed performance enhancing techniques can be roughly categorized into link layer solutions (see for example [Parsa and Garcia-Luna-Aceves, 1999a; Parsa and Garcia-Luna-Aceves, 1999b]), end-to-end solutions where the TCP end-to-end semantic is maintained (see for example [Ratnam and Matta, 1998b; Ratnam and Matta, 1998a; Cáceres and Iftode, 1995; Vaidya et al., 1999; Biaz and Vaidya, 1999; Balakrishnan et al., 1995; Balakrishnan et al., 1997b; Badrinath and Sudame, 1996]) and non end-to-end solutions where the end-to-end semantic is violated in one form or another (see for example [Bharadwaj, 1999; Bakre and Badrinath, 1995; Bakre and Badrinath, 1997]). Various link layer and end-to-end approaches can be quite effective for connections over wireless links through improved error correction, local retransmission and distinguishing congestion losses from link failure losses for TCP. In a connection that traverses a satellite link on the other hand, the main bottleneck in TCP performance is the large delay-bandwidth product nature of the satellite link (an overview of research on TCP over satellite can be found in for example [Allman et al., 2000]). Over such a link the normal TCP window dynamics results in significant latency before the channel is fully utilized. This problem cannot be effectively solved simply by improving the satellite channel quality, or by using large initial window size end-to-end. This is because a connection using the satellite link typically also has a terrestrial part, thus using large window end-to-end could affect the performance and fairness of the terrestrial part of the connection.

One typical non end-to-end solution that has been adopted by many satellite data communication service providers (see the previous chapter) is the TCP connection splitting/spoofing technique. The idea behind this technique is to segregate the end-to-end connection into segments so that each can be optimized separately. In particular the TCP window over the satellite segment can be opened up faster. This involves placing at the Network Operating Center (NOC) a splitting proxy that acknowledges end user packets on behalf of the remote server and acknowledges the remote server on behalf of the end user (assuming the end user is directly connected to the satellite down link). This proxy operates at the transport level and therefore breaks the end-to-end semantic of TCP, which is addressed in the previous chapter. The benefit however, is that

1 Splitting a connection results in shortened TCP feedback control loops, so that the remote server receives ACKs much sooner from the NOC (proxy) than from the end user, and therefore its window can quickly ramp up. At the same time, packet retransmission and loss recovery are also localized to be within each segment;

2 By segregating an end-to-end connection into segments, each segment can be enhanced separately. This is essential because different link characteristics may require very different enhancement schemes, which would not be appropriate if applied end-to-end. In particular, over the satellite link larger initial window size can be used. Furthermore, certain faster non-TCP based transmission rate control schemes can also be used.

Another example of a proxy that also shortens the TCP feedback control loop is a common web cache. When there is a "hit" at the cache, the file is directly sent to the client from the cache. When there is a "miss", the cache opens up a connection to the remote server and starts downloading the file to the cache (for cacheable objects), while forwarding packets to the client at the same time. Thus the cache automatically "breaks" the server-client transfer into two separate connections [Rodriguez et al., 2000]. In terms of TCP performance of the file transfer, this has exactly the same effect as split TCP (although the connection establishment is different). However, in this case the TCP semantic is preserved because the cache does not spoof the client's address, and so it acknowledges the server on behalf of itself rather than "pre-ack" on behalf of the client [2].

Figures 8.1 and 8.2 illustrate the packet flow between a client and a server with a splitting proxy and a cache (upon cache hit) in the middle, respectively. In the case of a splitting proxy the initial connection establishment (three-way handshake) and connection closing (which is not shown here) are usually done in an end-to-end fashion, although there are systems that split the connection establishment part as well. During the data transfer period, the connection is split in two, meaning the proxy acknowledges to the server, the client acknowledges to the proxy, and the proxy relays packets from the server to the client. Same procedure is followed in the other direction. The two connections are inevitably coupled, since the proxy cannot forward onto one connection any data packet it has not received from the other connection. However, they keep separate queues and sequence numbers. Also the proxy does not relay out-of-order packets from one to the other thus it acts as a virtual host of the session. With a cache there are two separate connection from the very beginning, i.e., three-way handshake is first conducted between the client and the cache, and if there is a miss, another three-way handshake is conducted between the cache and the server, as shown in Figure 8.2. Both situations result in approximately the same delay in connection establishment for a single connection.

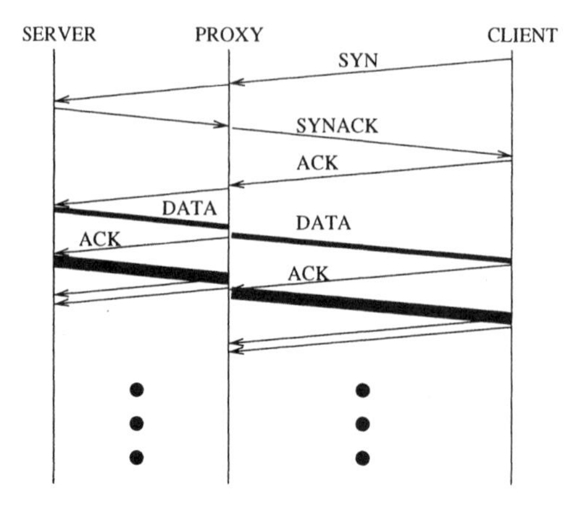


*Figure 8.1.* File transfer using a splitting proxy

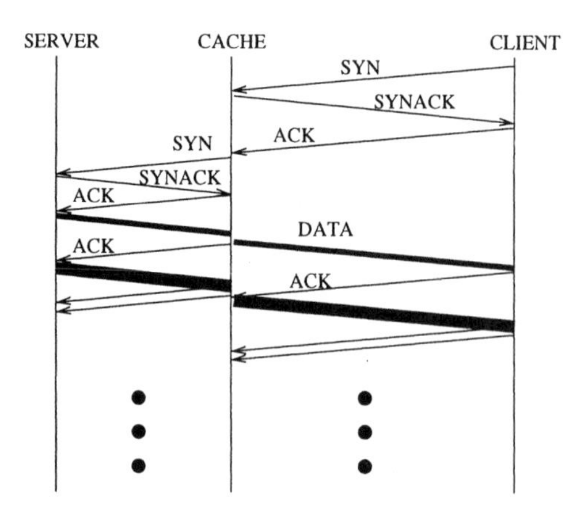


*Figure 8.2.* File transfer using a cache upon miss

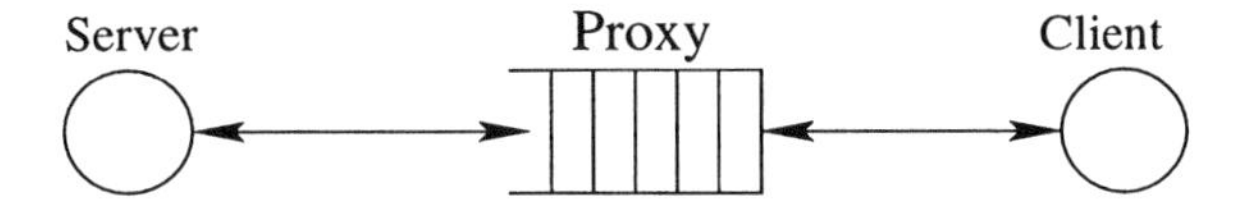


*Figure 8.3.* Network Model

In a typical satellite system TCP splitting proxy and web caches are often both used to enhance end-to-end performance. TCP splitting is used to speed up transmission over the satellite link and increase utilization of the channel, while caching is used to reduce latency by pushing the content closer to end users. They are often co-located at the NOC. Since both enhancements result in breaking up the end-to-end packet flow (either at the transport layer or at the application layer), interesting performance implications arise as we will see in subsequent discussion. In the next few sections, we will first present a few simple mathematical models to derive the TCP performance (mainly latency) when a splitting proxy is used between a server and a client. These models are derived for a general purpose of studying the effect of a splitting proxy. We will then conduct a sequence of experiments over a specific commercial satellite system with combined use of spoofing proxy and cache. Based on these results we will also discuss a sequence of system level design issues.

The performance gain of using a splitting proxy has been studied via both simulation and experiment. Interested readers are referred to for example [Ishac and Allman, 2001; Bharadwaj, 1999; Bakre and Badrinath, 1995; Bakre and Badrinath, 1997; Ehsan et al., 2002]. For quantitative analysis see for example [Rodriguez et al., 2000; Liu and Ehsan, 2002].

# 2. MODEL-BASED ANALYSIS

## 2.1 Network Model

Our analysis is based on a two-link model with one end host on each side and a proxy in the middle, as shown in Figure 8.3. This is a simple abstraction of a possibly more general connectivity. In reality each of the two links may contain multiple intermediate routers and physical links, but are abstracted into a single link with a single round-trip time (RTT) parameter and a single loss rate parameter. Note that under such abstraction the transmission rate of the server essentially represents the slowest physical link on the path from the server to the client. File transfer is our main application of interest in this study. Without loss of generality file transfers are considered to be from the server to the client.

The main performance metric we study in this section is file transfer latency, compared between using and not using a proxy. As shown in Figures 8.1, the connection establishment results in approximately the same amount of delay whether the proxy is used or not. We therefore do no include this initial delay in our analysis and instead solely focus on the delay of data transfer, which is the duration between when the server sends the first data packet of a file and the time when the client receives the last data packet of a file. As discussed earlier since the data transfer stages are virtually identical in the case of a splitting proxy and the case of a cache miss, subsequent analysis would apply to both cases [3].

We will assume that a file contains exactly $M$ segments of the maximum segment size (MSS). This is an approximation to an arbitrary file size whose last segment may be a fraction of MSS. However, this does not affect our method of analysis, and also does not affect the comparison between with and without the proxy. We assume that both the end-to-end connection (server-client) and the split connections (server-proxy and proxy-client) have the same value of slow start threshold (ssthresh), $W_{ss}$, and the maximum window size $W_{max}$. These two values are also assumed to be in number of segments rather than number of bytes to simplify the analysis.

The server, the proxy and the client each has a transmission rate of $C_1$, $C_p$, and $C_2$, respectively. Assuming packet length of $D = MSS + 40$ (including both TCP and IP headers), the time it takes for the server to transmit a packet is $\mu_1 = \frac{D}{C_1}$, and $\mu_p = \frac{D}{C_p}$, $\mu_2 = \frac{D}{C_2}$ for the proxy and the client, respectively. When there are two separate connections, we assume a per-packet processing delay of $t_p$ at the proxy. All other processing delays are ignored. We assume that each link has the same propagation delay and transmission rate in both directions. The one-way propagation delay on the server-proxy link and the proxy-client link are denoted by $I_1$ and $I_2$, respectively. Both links are assumed to be symmetric in terms of round trip time [4]. Throughout our analysis, we assume that the transmission time of an ACK is negligible.

We further assume that the TCP sender is only constrained by the congestion window and not the advertised receive window size. Most work in TCP analysis assumes an infinite source, e.g., [Cardwell et al., 2000; Padhye et al., 2000; Lakshman and Madhow, 1997]. However, when we have two connections, the window of the second connection (proxy-client) evolves not only according to the window dynamics of TCP, but also according to the availability of packets (from the server-proxy connection). For example, the first connection may not "catch up" with the second connection due to factors like initial window size, transmission rate, etc.. Therefore the window of the second connection will be forced to grow at a slower rate. We will discuss this in this section.

## 2.2 Lossless Links

Let us first assume that both links are free of random losses and congestion loss's. Assuming that the window grows in the slow start and congestion avoidance stages until the maximum window size is achieved, the number of windows that is needed to cover a file of $N$ segments can be calculated by extending the method presented in [Kurose and Rose, 2000]. We also assume that delayed ACK is implemented. As shown in [Allman and Paxson, 1999], since one ACK is generated for every $b$ packets received before the timer expires, the rate of exponential growth of the congestion window is $r = 1 + \frac{1}{b}$, which equals 2 when no delayed ACK is used. Let $w_o$ denote the initial window size. Let $M_s$ be such that

$$w_o r^{S-1} < W_{ss} \leq w_o r^S, \tag{8.1}$$

if $N > \sum_{i=1}^{M_s} w_o r^{i-1}$, i.e., the slow start threshold *ssthresh* is reached during the $(M_s + 1)^{th}$ window given the file is big enough. Therefore the $(M_s + 1)^{th}$ window size is $W_{ss}$ and the $(M_s + 2)^{th}$ window size is $W_{ss} + 1$, and so on. Similarly, let $M_x$ be such that

$$W_{ss} + \frac{M_x - M_s - 1}{b} < W_{max} \leq W_{ss} + \frac{M_x - M_s}{b}, \tag{8.2}$$

i.e., the maximum window size is achieved during the $(M_x + 1)^{th}$ window if the file is big enough. All subsequent windows have the same window size of $W_{max}$. The number of windows needed to transfer a file is then given by the following:

$$K = \begin{cases} \min\{k : \sum_{i=1}^{k} w_o r^{i-1} \geq N\} \text{ if } k \leq M_s \\ \min\{k : \sum_{i=1}^{M_s} w_o r^{i-1} + \sum_{i=M_s+1}^{k} (W_{ss} + \frac{i-M_s-1}{b}) \geq N\} \\ \qquad\qquad\qquad\qquad \text{if } M_s < k \leq M_x \\ \min\{k : \sum_{i=1}^{M_s} w_o r^{i-1} + \sum_{i=M_s+1}^{M_x} (W_{ss} + \frac{i-M_s-1}{b}) \\ \qquad + \sum_{i=M_x+1}^{k} W_{max} \geq N\} \text{ if } M_x < k \end{cases}. \tag{8.3}$$

**2.2.1 Delay Models.** We first consider an end-to-end connection between the server and the client, i.e., assuming that the proxy is simply acting as a normal router. Assuming that the links are lossless and that connections are only constrained by congestion window size, after the server sends a window's packets it waits for the first ACK to come back if it takes longer for the ACK to arrive than it takes to transmit the window's worth of data. The time it takes to transmit the $k^{th}$ window is a function of the packet transmission time at the sender given by

$$t_k(\mu_1) = \begin{cases} w_o r^{k-1} \mu_1 & \text{if } k \leq M_s \\ (W_{ss} + \frac{k-M_s-1}{b})\mu_1 & \text{if } M_s < k \leq M_x \\ W_{max}\mu_1 & \text{if } M_x < k \end{cases}. \tag{8.4}$$

Therefore if $\mu_1 \geq \mu_p$, which indicates that the proxy transmits at least as fast as the server and thus packets will not experience queuing delay at the proxy, the round-trip time of the end-to-end connection is $2(I_1 + I_2)$. We define $R_e$ to be the time it takes for the first ACK to arrive after the first packet was sent, thus

$$R_e = \mu_1 + \mu_p + 2(I_1 + I_2) + (b-1)\mu_1 = b\mu_1 + \mu_p + 2(I_1 + I_2). \quad (8.5)$$

Note that this expression assumes that there are at least $b$ packets in a window so that the receiver can immediately return an ACK upon receipt of the $b^{th}$ packet. If for example $w_o = 1$ and $b = 2$, then the receiver may have to wait for the delayed ACK timer to expired to return an ACK. In the rest of our analysis we will ignore this difference, which can be easily taken into account. The total time it takes to transfer the file is then

$$T_e(N) = N\mu_1 + \sum_{k=1}^{K-1} [R_e - t_k(\mu_1)]^+ + I_1 + I_2 + \mu_p, \quad (8.6)$$

where $[a]^+ = a$ for $a$ positive and 0 otherwise. This latency reflects the total transmission time, the time that the server spends waiting for ACKs, and the time for the last window to reached the client.

When $\mu_1 < \mu_p$, packets could build up at the proxy waiting to be transmitted into the slower link and experience additional queuing delay at the proxy. In this case the ACKs of the same window arrive at the server approximately $\mu_p$ apart instead of $\mu_1$, thus the server may need to wait for every ACK of the same window instead of stalling after sending out the entire window. The above stall time analysis therefore does not directly apply. We derive the latency by examining from the client's side. Since $\mu_p > \mu_1$, the client receives packets of the same window continuously at rate $1/\mu_p$. The time that the client is idle is therefore $[\mu_1 + \mu_p + 2(I_1 + I_2) + (b-1)\mu_p - t_k(\mu_p)]^+$, where $t_k(\mu_p)$ is the time it takes the client to receive the $k^{th}$ window, and $t_k(\cdot)$ is given by the following:

$$t_k = \begin{cases} w_o 2^{k-1}\mu_p & \text{if } k \le M_s \\ (W_{ss} + k - M_s - 1)\mu_p & \text{if } M_s < k \le M_x \\ W_{max}\mu_p & \text{if } M_x < k \end{cases} \quad (8.7)$$

The latency is then

$$\begin{aligned} T_e(N) &= N\mu_p + I_1 + I_2 + \mu_1 \\ &\quad + \sum_{k=1}^{K-1} [\mu_1 + 2(I_1 + I_2) + b\mu_p - t_k(\mu_p)]^+ \end{aligned} \quad (8.8)$$

which reflects the time the client spends receiving the file, waiting for the next window, and the time for the first window to reach the client. Redefining

$$R_e = \min(\mu_1, \mu_p) + b\max(\mu_1, \mu_p) + 2(I_1 + I_2), \quad (8.9)$$

(8.6) and (8.7) can be combined into the following:

$$T_e(N) = N\max(\mu_1, \mu_p) + I_1 + I_2 + \min(\mu_p, \mu_1) + \sum_{k=1}^{K-1} [R_e - t_k(\max(\mu_1, \mu_p))]^+. \quad (8.10)$$

When the proxy is used, we have two serial connections. Note that these two connections are not independent but coupled by data. This is because the second connection (proxy-client) cannot send any data packets it has not received from the first connection (server-proxy) and therefore be constrained. This can be caused by a much larger initial window size and/or a much shorter round-trip time on the second connection. In this scenario the second connection has a limited source based on the sending of the first connection. In [Liu and Ehsan, 2001] we developed a rather detailed and complicated model for this scenario. However, similar qualitative insight can be obtained without having to go through the detailed analysis. Therefore for the rest of our discussion we will assume that the second connection is never constrained by the first connection, which could imply $w_o \geq w_o^{'}$, $\mu_1 \leq \mu_p$ and/or $I_1 \leq I_2$, where $w_o^{'}$ is the proxy's initial window size. Assuming both connections use the same initial window size, $ssthresh$ and $W_{max}$, we get $M_s$, $M_x$, and $K$ as shown before for both connections. The proxy receives the first packet from the server at time $\mu_1 + I_1$. Due to $t_p$ delay for processing at the proxy, the proxy starts sending this packet to the client at time $\mu_1 + I_1 + t_p$. From this point on, we only need to focus on the second connection since the latency is only determined by this connection. By following the same analysis, we have the total latency for the proxy case

$$T_p(N) = \mu_1 + I_1 + t_p + N\mu_p + \sum_{k=1}^{K^{'}-1} [R_2 - t_k(\mu_p)]^+ + I_2, \quad (8.11)$$

where $t_k(\cdot)$ is given in (8.7), $K^{'}$ is the total number of windows needed for the transfer, and

$$R_2 = \mu_p + 2I_2 + (b-1)\mu_p = b\mu_p + 2I_2 \quad (8.12)$$

is the time it takes for the ACK to come back to the proxy. This latency reflects the initial delay for the first packet to arrive at the proxy, the total transmission time at the proxy, stall time and the time for the last packet to reach the client.

Figure 8.4 compares the numerical results from our model with *NS2* simulation, for both the end-to-end and proxy schemes. In this case, $C_1 = C_p = C_2 =$ 1Mbps. The initial window size is set to 1 and 4, respectively. Unless pointed out explicitly, our numerical results and simulation throughout this paper are

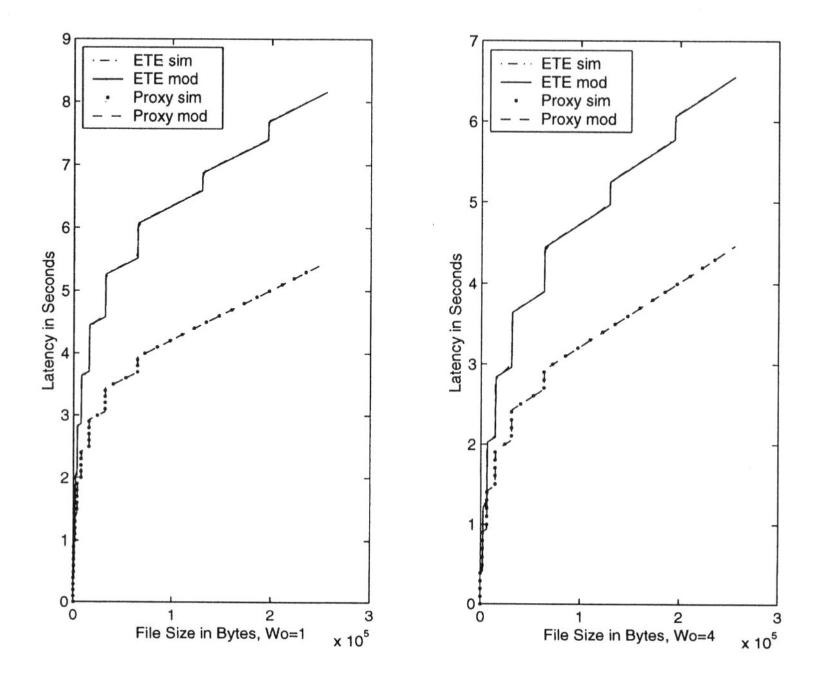


*Figure 8.4.* Latency vs. file sizes, with initial window size of 1 and 4, respectively.

based on the following parameters: MSS=512 bytes, ssthresh=128 segments. In both cases, $I_1 = 150$ ms, $I_2 = 250$ ms. Each graph contains four curves, two from NS simulation ("sim") and two from our model ("mod"). We see that each pair ("sim" and "mod") overlaps almost completely.

## 2.3 Links with Random Losses

When losses (either due to congestion or link failure) are present the analysis becomes more complicated. Moreover the analysis is largely limited to the steady-state study of TCP connections, which is applicable in the case of an unlimited file transfer, less accurate in the case of a finite TCP connection, and much less in the case of a proxy as we will show.

**2.3.1 The Server-Proxy Link is lossless.** If we assume that the server-proxy is lossless, then the methods introduced in [Lakshman and Madhow, 1997; Padhye et al., 2000; Cardwell et al., 2000] can be applied to determine the throughput and delay of the proxy-client connection. In particular, the TCP bulk data transfer throughput is shown to be well approximated by

$$\lambda(RTT, p) = \sqrt{\frac{3}{2bp}}\frac{1}{RTT}, \tag{8.13}$$

where $p$ is the probability of loss for a single packet at low loss rate [Lakshman and Madhow, 1997], and is more accurately approximated in [Padhye et al., 2000] by considering timeouts. These results were developed for bulk TCP transfers and were based only on analyzing the TCP congestion avoidance phase. In [Cardwell et al., 2000] it was shown that they can be reasonably effective when applied to short TCP connections if combined with slow start analysis.

The performance implication of using a proxy when losses are present immediately follows. The throughput of a TCP transfer is inversely proportional to the connection round-trip time and the square-root of the loss rate. If losses are concentrated on the proxy-client link, then using the proxy effectively isolates the part of the connection that involves loss, and reduces the round-trip time required to recover the losses, thus achieves higher throughput and lower latency. The same key concept can be seen in schemes like Snoop TCP [Balakrishnan et al., 1995; Balakrishnan et al., 1997b], WTCP [Ratnam and Matta, 1998b; Ratnam and Matta, 1998a], and [Parsa and Garcia-Luna-Aceves, 1999a; Parsa and Garcia-Luna-Aceves, 1999b] that use local retransmission (some at the link layer, some at the transport layer).

Specifically, denoting the loss rate on the proxy-client link by $p_2$, and the throughput by $\lambda(RTT, p_2)$, the transfer latency of a file of size $N$ using end-to-end connection is given by

$$\begin{aligned} T_e &= \sum_{n=0}^{N} p(n)(T_e(n) + \frac{N-n}{\lambda(RTT, p_2)}) \\ &= \sum_{n=1}^{N} p(n)T_e(n) + \frac{N - m_{loss}}{\lambda(2(I_1 + I_2), p_2)} \end{aligned} \quad (8.14)$$

where $p(n)$ is the probability that $n$ packets are successfully sent before the first loss occurs:

$$p(n) = \begin{cases} (1-p)^n p & n < N \\ (1-p)^N & n = N \end{cases} \quad (8.15)$$

$m_{loss}$ is the expected number of packets sent before the first loss occurs:

$$m_{loss} = \frac{(1-(1-p_2)^N)(1-p_2)}{p_2} + 1. \quad (8.16)$$

$T_e(\cdot)$ is the latency function of an end-to-end connection shown earlier.

Assuming that the proxy-client connection is not constrained by the server-proxy connection (e.g., $I_1 < I_2$, $\mu_1 < \mu_2$),

$$T_p = \sum_{n=1}^{M} p(n)T_p(n) + \frac{N - m_{loss}}{\lambda(2I_2, p_2)}. \quad (8.17)$$

The difference of the two is then

$$T_e - T_p = \sum_{n=1}^{N} p(n)(T_e(n) - T_p(n)) + \frac{(N - m_{loss})\sqrt{2bp_2}}{\sqrt{3}}(2I_1)$$
$$\approx (\sum_{n=1}^{N} p(n)(k_n - 1) + \frac{(N - m_{loss})\sqrt{2bp_2}}{\sqrt{3}})(2I_1), \quad (8.18)$$

where $k_n$ is the number of window needed to cover a file of $n$ segments. The approximation in the second equation is based on the assumption that the link capacity is not filled during the transfer of $n$ packets. For a given file size and loss rate on the proxy-client link, the first term of the above equation is a constant, and the amount of gain in using the proxy depends on the round-trip time of the server-proxy connection.

Figure 8.5 compares the latency obtained using this analysis with that of simulation. There is an obvious discrepancy between the two curves. This is mainly due to the fact that the delay model assumes that the connection goes into steady-state right after the first loss. The two curves eventually approach each other as the file size increases (number of packets sent in this figure). This is because for a large file transfer the effect of the above assumption is diluted (since the effect of steady state will dominate).

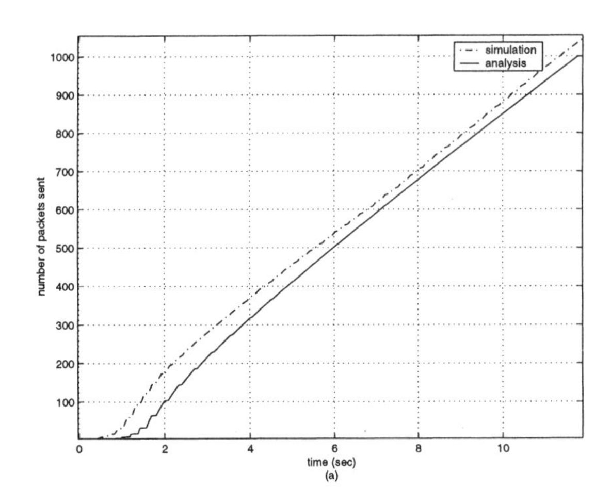


*Figure 8.5.* Latency when splitting is used and the first link is lossless

**2.3.2 Random Losses on Both Links.** When losses are present on both links, using the previous analysis for both connections provides only gross approximation. Suppose the loss rates on the server-proxy link and the proxy-client link are $p_1$ and $p_2$, respectively. Assuming losses are independent, the overall loss rate experienced by an end-to-end connection is $p = p_1 + p_2 - p_1 p_2$. For an infinite file transfer, in the long run the server-proxy connection has an average throughput $\lambda_1 = \sqrt{\frac{3}{2bp_1}}\frac{1}{R_1}$ and the proxy-client connection has an average throughput $\lambda_2 = \sqrt{\frac{3}{2bp_2}}\frac{1}{R_2}$ if unconstrained by the server-proxy connection. The slower one of the two is going to dominate the combined throughput and delay. However, both values are greater than the throughput of the end-to-end connection $\lambda = \sqrt{\frac{3}{2bp}}\frac{1}{R_1+R_2}$ since it has a larger loss rate and a larger round-trip time. Therefore by segregating the server-client connection into parts so that each has a smaller loss rate and round-trip time, using the proxy achieves higher throughput and thus lower latency.

For a finite file transfer, the latency of the end-to-end connection is given by (8.14) with loss rate $p$. When using the proxy, we consider $m_1, m_2$, the expected number of packets sent successfully before the first packet loss occurs to the two connections, respectively. The steady-state bulk transfer throughput $\lambda_1$ and $\lambda_2$ can be derived as before. Following this we can characterize the sending process and congestion window evolution of the server for the first $m_1$ packets, and use $\lambda_1$ as an approximation to the remaining of the transfer. The characterization of the proxy can be obtained by listing all possible cases comparing $m_1$ and $m_2$, $T_e(m_1)$ and $T_p(m_1)$, $\lambda_1$ and $\lambda_2$. However this method does not provide reasonably accurate estimate. A better model is likely to be found by describing the state transitions of both the server and the proxy through Markov Chain analysis.

## 3. DISCUSSION

Due to the complexity involved, detailed analysis is only presented for the lossless scenario in the previous section, and our discussion in the lossy scenario is largely limited to the steady-state behavior. In this section we will focus on the model developed for the lossless scenario to obtain useful insight through such analysis, especially for short file transfers.

### 3.1 Initial Window Size

From the definition of $M_s$ and $M_x$, $M_s = \lceil \log_r(\frac{W_{ss}}{w_o}) \rceil$, $M_x = b(W_{max} - W_{ss}) + M_s$. Consider a file that finishes transferring during the slow start phase, the total number of windows needed to cover the file, $K$, would be such that $K \leq M_s$. Since $N \leq \sum_{i=1}^{K} w_o r^{i-1} = w_o \frac{r^K - 1}{r-1}$, this means $N \leq \frac{W_{ss} - w_o}{r-1}$. $K$ is the smallest integer that satisfies $N \leq w_o \frac{r^K-1}{r-1}$, therefore we have

$$K = \lceil \log_r(\frac{N}{w_o}(r-1)+1) \rceil$$

$$\approx \log_r(\frac{N}{w_o}(r-1)+1)+1. \quad (8.19)$$

Assume $\mu_1 W_{ss} \leq R_e$, i.e., the link (or pipe capacity) is not filled during slow start, and $\mu_1 < \mu_p$ so that the proxy is slower than the server, we then have

$$\begin{aligned} T_e &= N\mu_p + \sum_{k=1}^{K-1}[R_e - t_k(\mu_p))]^+ + I_1 + I_2 + \min(\mu_1, \mu_p) \\ &= N\mu_p + (K-1)R_e - \sum_{k=1}^{K-1} w_o r^{k-1}\mu_p + I_1 + I_2 + \mu_1 \\ &= N\mu_p + (K-1)R_e - w_o\frac{r^{K-1}-1}{r-1}\mu_p + I_1 + I_2 + \mu_1 \\ &\approx R_e \log_r(\frac{N}{w_o}(r-1)+1) + C_1, \end{aligned} \quad (8.20)$$

where $C_1 = I_1 + I_2 + \mu_1$. This last equation is the same as presented in [Cardwell et al., 2000], but derived in a different way here.

Similarly, for an initial window size $w_o^{'}$ used by the proxy, assuming $\mu_p W_{ss} \leq R_2$, we have

$$\begin{aligned} K^{'} &= \lceil \log_r(\frac{N}{w_o^{'}}(r-1)+1) \rceil \\ &\approx \log_r(\frac{N}{w_o^{'}}(r-1)+1)+1 \end{aligned} \quad (8.21)$$

and

$$\begin{aligned} T_p &= N\mu_p + \sum_{k=1}^{K^{'}-1}[R_2 - t_k(\mu_p)]^+ + I_1 + I_2 + \mu_1 + t_p \\ &= N\mu_p + (K^{'}-1)R_2 - w_o^{'}\frac{r^{K^{'}-1}-1}{r-1}\mu_p + I_1 + I_2 + \mu_1 + t_p \\ &\approx N\mu_p + R_2\log_r(\frac{N}{w_o^{'}}(r-1)+1) - N\mu_p + I_1 + I_2 + \min(\mu_1, \mu_p) \\ &\approx R_2 \log_r(\frac{N}{w_o^{'}}(r-1)+1) + C_2, \end{aligned} \quad (8.22)$$

where $C_2 = I_1 + I_2 + \mu_1 + t_p$. Note that $C_1 \approx C_2$ and both are close to one half of $R_e$. Since $R_e > R_2$, $w_o$ has to be greater than $w_o^{'}$ in order to achieve the same delay. More specifically,

$$(\frac{N}{w_o}(r-1)+1)^{R_e} \approx (\frac{N}{w_o^{'}}(r-1)+1)^{R_2}, \quad (8.23)$$

which leads to

$$w_o = \frac{N(r-1)}{(\frac{N}{w_o'}(r-1)+1)^{R_2/R_e}-1}. \tag{8.24}$$

Table 8.1 shows some values of $N$ and $w_o$ based on this approximation for $r = 2, w_o' = 1, R_2/R_e = 0.5$ and $W_{ss} = 128$ segments. We see that in order to achieve similar latency even for relatively small files we need significantly larger initial window size for the end-to-end connection.

*Table 8.1.* Initial window size of the end-to-end connection

| File Size (Kbytes) | 10 | 20 | 25 | 30 | 35 |
|---|---|---|---|---|---|
| $w_o$ | 5 | 7 | 8 | 9 | 9 |

## 3.2 Slow or Congested Proxy

When the same initial window size $w_o$ is used, the difference in delay between the two is

$$\begin{aligned} T_e - T_p &= (R_e - R_2)\log_r(\frac{N}{w_o}(r-1)-1) \\ &= (2I_1 + \mu_1)\log_r(\frac{N}{w_o}(r-1)-1) \end{aligned} \tag{8.25}$$

This difference increases as $N$ and $I_1$ increase, but seems invariant to changes in $\mu_p$. As $\mu_p$ increases, which corresponds to a slower proxy, the difference in delay remains constant so long as $\mu_p w_o r^{k-1} \leq R_2$ for any $k \leq K$. However, as $\mu_p$ keeps increasing to the point where the pipe is filled before the file transfer completes, the difference quickly reduces. In particular, if the pipe capacity is achieved during the $k_e^{th}$ window for the end-to-end connection, i.e., $\mu_p w_o r^{k_e - 1} \geq R_e$, then $k_e = \lceil \log_r(\frac{R_e}{\mu_p w_o}) + 1 \rceil$ and

$$\begin{aligned} T_e &= N\mu_p + R_e(k_e - 1) - \sum_{k=1}^{k_e - 1} w_o r^{k-1}\mu_p + I_1 + I_2 + \mu_1 \\ &\approx N\mu_p + R_e \log_r(\frac{R_e}{\mu_p w_o}) - \frac{R_e - \mu_p w_o}{r-1} + I_1 + I_2 + \mu_1. \end{aligned} \tag{8.26}$$

$$\tag{8.27}$$

We can obtain a similar expression for $T_p$:

$$\begin{aligned} T_p &\approx N\mu_p + R_2 \log_r(\frac{R_2}{\mu_p w_o'}) - \frac{R_2 - \mu_p w_o'}{r-1} \\ &\quad + I_1 + I_2 + \mu_1 + t_p. \end{aligned} \tag{8.28}$$

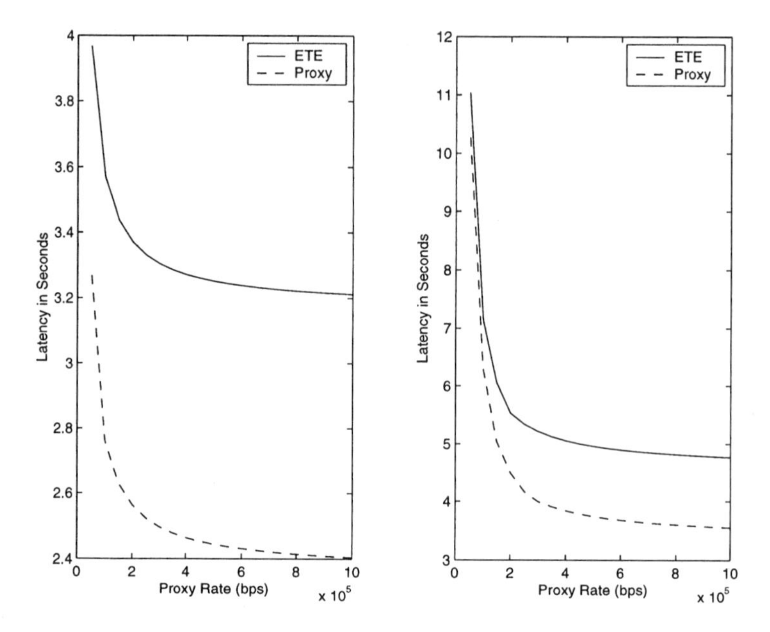


*Figure 8.6.* Latency vs. $C_p$, the transmission rate of the proxy. File size is 11 Kbytes and 51 Kbytes, for the graph on the left and right, respectively. $C_1 = C_2 = 1$ Mbps in both cases. These graphs are derived from our model.

Therefore

$$T_e - T_p \approx R_e \log_r(\frac{R_e}{\mu_p w_o}) - R_2 \log_r(\frac{R_2}{\mu_p w_o^{'}}) + \frac{\mu_p(w_o - w_o^{'}) - 2I_1 - \mu_1}{r-1}. \tag{8.29}$$

where $R_e = 2(I_1 + I_2) + \mu_1 + b\mu_p$ and $R_2 = 2I_2 + b\mu_p$. The above value decreases as $\mu_p$ increases. This result can be observed in Figure 8.6. A slower proxy (increased $\mu_p$ and decreased transmission rate) can be viewed as an approximation to a busier or more congested proxy, because under such a situation each TCP connection only gets a fraction of the total proxy capacity (assuming the proxy has sufficient buffer), and queuing is increased. This result shows that as the proxy becomes busy, the gain from using separate connections reduces because the bottleneck dominates the overall performance no matter which scheme we use. In a system where a proxy is placed at the aggregation point of incoming traffic, adequate provisioning of such a proxy becomes very important since otherwise very little is gained from using a proxy.

## 3.3 File Size

In the case where the file transfer enters the congestion avoidance stage, i.e., $N > \frac{W_{ss}-w_o}{r-1}$, similar analysis applies and we have

$$T_e = N\mu_p + R_e(k_e - 1) - \sum_{k=1}^{k_e-1} t_k(\mu_p) + I_1 + I_2 + \mu_1, \quad (8.30)$$

$$T_p = N\mu_p + R_2(k_e^{'} - 1) - \sum_{k=1}^{k_e-1} t_k(\mu_p) + I_1 + I_2 + \mu_1 + t_p, \quad (8.31)$$

where $k_e$ and $k_e^{'}$ are the total number of window sent before the pipe becomes full for the end-to-end and the split connection, respectively. In case when the file finishes transfer before the pipe is full, $k_e = k_e^{'} = K$, and the difference between the two is mainly $(R_e - R_2)(K - 1)$, which increases as $N$ increases ($K$ increases with $N$). However, if $N$ is large enough and the pipe is filled up before the transfer completes, then the difference between the two stays constant, and both increase with rate $\mu_p$ as the file size increases. This can be observed in Figure 8.7.

## 3.4 Connection With Asymmetric Segments

Suppose $w_o \leq w_o^{'}$ and $R_1 > R_2$, i.e., the proxy-client connection is constrained by the server-proxy connection throughout the file transfer. Further assume that the file transfer is only limited to slow start phase. Following our previous analysis, we can show

$$\begin{aligned} T_p(N) &= R_1 \log_r(\frac{N(r-1)}{w_o} + 1) - I_1 + I_2 \\ &\approx R_1 \log_r(\frac{N(r-1)}{w_o} + 1), \end{aligned} \quad (8.32)$$

with an error within half of $R_1$.

Suppose we now let $R_1 < R_2$, but keep $R_1 + R_2$ unchanged, and let $w_o > w_o^{'}$, then using our earlier analysis we get

$$T_p(N) = R_2 \log_r(\frac{N(r-1)}{w_o^{'}} + 1) \quad (8.33)$$

while the latency of end-to-end connection remain the same $T_e(N) = (R_1 + R_2)\log_r(\frac{N(r-1)}{w_o} + 1)$. We see that when using the proxy, the longer connection of the two ($\max\{R_1, R_2\}$) determines the total latency. As the difference between the two round-trip times increases, the gain from using the proxy reduces. Under this scenario the performance of the proxy is maximized when

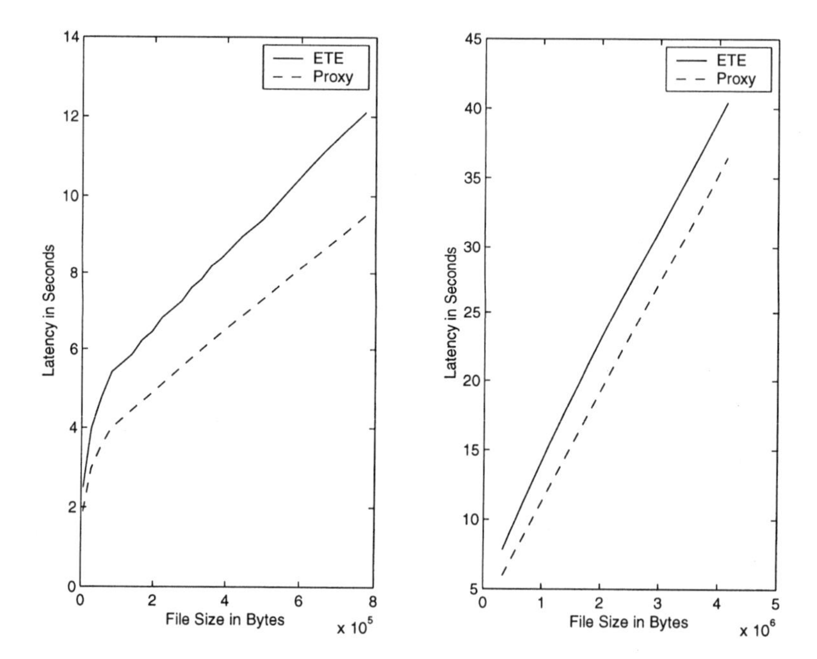


*Figure 8.7.* Latency vs. file size. For small files the latency of an end-to-end connection increases faster than that of split connections. However as file size grow big enough to fill up the capacity, the two have same growth rate and the difference stays constant. These graphs are derived from our model.

the two connections are "similar" – $R_1 \approx R_2, w_o \approx w_o^{'}$. Same argument can also be derived for file transfers that enter congestion avoidance phase. This is an interesting observation since such proxies are commonly used in a heterogeneous environment where links have very different properties (e.g., in the satellite communication the satellite link has a much longer propagation delay than the terrestrial link). The results shown here indicate that while it is very important to optimize each link separately, it is equally important to minimize the asymmetry between links. If separate optimizations only increases the difference, e.g., making the fast link even faster, the resulting performance might not be improved. We will see an example of this effect in the measurement study.

# 4. THE EXPERIMENT SYSTEM

In this section and the next we describe a measurement-based study conducted over a commercial satellite system that uses a geo-stationary (GEO) satellite (Ku band) for forward data transfer (from the NOC to the client/end host) and a regular phone line as the return channel (from the client/end host to the NOC via an ISP), as depicted in Figure 7.2. This is a specific example of using split connection to enhance the performance of TCP over satellite. We will present some interesting observations and will be able to use our more general analysis presented earlier to interpret the data.

Available bandwidth on the forward channel is up to 24 Mbps and the one way propagation delay of the satellite link is roughly 250 ms (however due to NOC configuration the maximum throughput we were able to achieve per client was 300-400 Kbytes/sec). The return link has 4 KHz bandwidth. The TCP connection splitting proxy (which we will simply call proxy in the following) is implemented on a Hybrid Gateway (HGW) located at the NOC. When the proxy is disabled, a packet from the server to the end host passes through the HGW as if passing through a normal router, and goes directly to a Satellite Gateway (SGW) connected to the satellite uplink. When the proxy is enabled , it breaks up the server-client end-to-end connection in two, and pre-acknowledges one on behalf of the other in the following way, as illustrated in Figure 8.1.

The web caches are also located at the NOC. Regardless of whether the splitting proxy is enabled or disabled, when caching is allowed an HTTP request received at the NOC first goes through the cache. For cacheable content, if a fresh copy of the requested file is located in the cache (a *hit*), the file is delivered to the client directly from the cache without going to the remote server. If the requested file is not found in the cache (a *miss*), the cache will open up a connection to the remote server to fetch the file, as shown in Figure 8.2. As we have discussed earlier, except for the connection establishment process, the data transfer essentially proceeds in an identical manner as that in connection splitting. Consequently the splitting proxy together with the cache result in an end-to-end connection split twice upon a cache miss and once upon a cache hit, as shown in Figures 8.8(a) and 8.8(b), respectively. Figures 8.8(c) and 8.8(d) illustrates cache miss and cache hit, respectively, when the proxy is disabled. Figures 8.8(e) and 8.8(f) illustrates the cases where connections bypass the cache (e.g., for contents not cacheable) with the splitting proxy enabled and disabled, respectively.

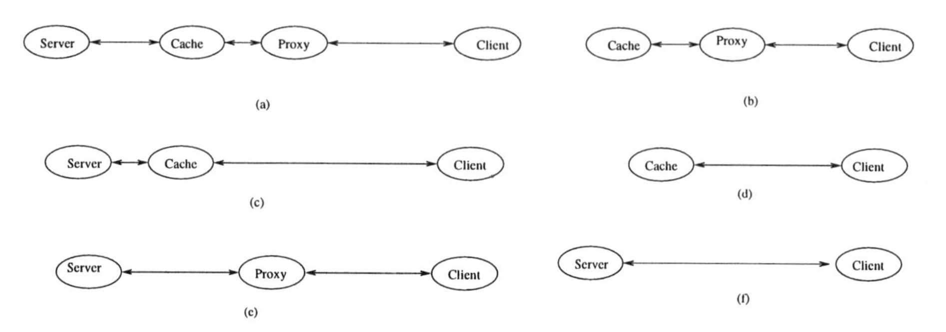


*Figure 8.8.* Experiment scenarios (a) Cache miss, proxy enabled; (b) Cache hit, proxy enabled; (c) Cache miss, proxy disabled; (d) Cache hit, proxy disabled; (e) No cache, proxy enabled; (f) No cache, proxy disabled

Important parameters of our system are as follows. The client is running Windows ME that uses TCP SACK [Mathis et al., 1996] with delayed acknowledgments (one ACK for every two received packets). Our web server is running Linux Redhat 7.1. Because of the high asymmetry in link speeds between the forward and return paths, we also use ACK filtering [Balakrishnan et al., 1997a] at the client and send one out of every four ACKs. Thus each ACK received at the server (assuming no loss) in general represents eight packets. Byte counting instead of ACK counting is used at the server as well as the proxy, so that the lower rate of ACKs does not reduce the data sending rate. The splitting proxy uses an initial window size of 64 Kbytes for the proxy-client connection over the satellite whenever enabled.

Our study consists of six scenarios: splitting enabled or disabled with cache hit or cache miss and the option of whether to bypass the cache or not, as shown in Figure 8.8. Whether to use the file in the cache or not is controlled by a *no-cache pragma* [Berners-Lee et al., 1995; Fielding et al., 1997] set in the request header. When set, this parameter tells the cache not to use the cached copy even if there is one and to get the most updated version from the server. Whether the connection splitting proxy is used or not is controlled by the end hosts. We have two end hosts, one of which has the proxy option enabled and the other one has the option disabled. This option is then encapsulated in the header of packets sent to the NOC. For comparison purposes, we always run experiments on these two hosts simultaneously. We download files from our web server onto both hosts repeatedly for durations of 1-2 hours per measurement point (measurements over this period are averaged into one data point).

Our connections and experiment traffic go through a dedicated HGW and a dedicated splitting proxy that are not shared by other traffic. However our connections does share the cache access, the satellite gateway and the satellite transponder with other through traffic since this is a live commercial system. Such a setup results in both controllable and uncontrollable factors as we will point out when we discuss our results in the next section.

The performance metrics we use in this study are the file transfer latency (using HTTP) and throughput. We define latency seen by the client as the time between when the SYN request is sent and the time when FINACK is sent by the client. For files with multiple embedded objects, this is defined as the time between when the first SYN request is sent and the time when the last FINACK is sent. Throughput is defined as file size divided by latency. Files used in this study contain a set of text files of varying sizes and a set of HTML files with varying number of embedded objects. Often for comparison purposes we keep the total file transfer size (base page plus embedded objects) the same while changing the number of embedded objects and thus the size of each object.

## 5. MEASUREMENT-BASED ANALYSIS

For comparison purposes and better illustration of our results, we define the *Gain Of Splitting* (GoS) as

$$\text{GoS} = \frac{\text{Throughput}_{\text{splitting}} - \text{Throughput}_{\text{end-to-end}}}{\text{Throughput}_{\text{end-to-end}}}.$$

Thus larger GoS means higher throughput gain from using the splitting proxy. A negative GoS means that the end-to-end connection has a higher throughput (or smaller latency) than the split connection. We will frequently use this metric in subsequent discussions.

### 5.1 Effect of File Size and Caching

In this subsection we compare the performance under scenarios described in Figures 8.8(a)-8.8(d). Files are downloaded onto both the splitting enabled and the splitting disabled hosts repeatedly over a one-hour period. Figure 8.9 shows the GoS of five files ranging from 10 Kbytes to 120 Kbytes in the cache hit and the cache miss cases.

The first observation is that splitting provides better performance gain for larger files up to a certain point. This result is consistent with our earlier analysis. This is due to the fact that the time spent in TCP connection establishment has a larger portion in the overall latency for smaller files and this time is not

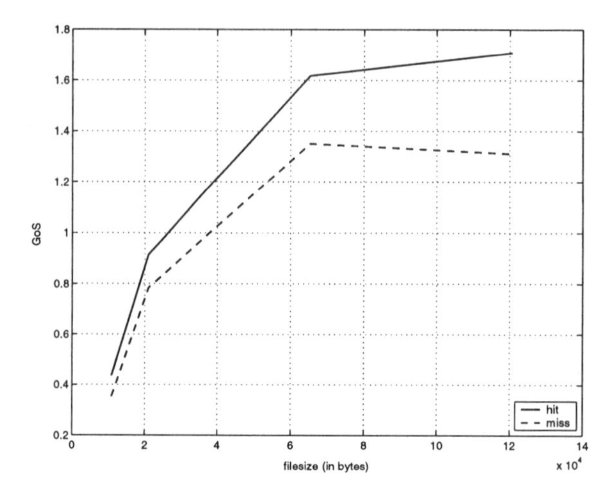


*Figure 8.9.* Comparing GoS in the case of cache hit and cache miss

reducible by using splitting. Furthermore, larger files benefit more from the large initial window size (for example a 1 Kbyte file and a 64 Kbyte file can both be transmitted in the first window and experience almost the same latency, if file transfer from the source to the proxy is fast enough). However, this gain saturates roughly around 80K. This is because when the satellite link is fully utilized (as in the case of a very large file), the difference in throughput gain between the two scenarios will start to reduce. The main benefit of using a split connection comes from sending faster over the satellite link. When the capacity of the satellite link is fully utilized this benefit starts decreasing (Analysis in Section 3 on file size indicates that as the file size increases, the difference in latency becomes constant, which means the difference in throughput decreases. Thus the gain goes down). The larger the file size, the more diluted this benefit gets. Therefore we expect GoS to go down as file size keeps increasing beyond 120 Kbytes (as a matter of fact it can already be seen in the cache miss case).

A second observation and a more interesting one from this comparison is that when a file is in the cache, the splitting gain is much higher. This can be more clearly seen in Figures 8.10 (a) and (b), where we compare the file transfer throughput separately for the cache hit and cache miss cases.

By comparing the two we see that the use of splitting at the hybrid gateway enhances the value of caching, i.e., when splitting is used, having the file in the cache provides significant increase in throughput over the case where the file has to be retrieved remotely from the server. In addition, this improvement increases as the file size increases (before capacity is reached). On the other hand, when splitting is disabled, whether the file is in the cache or not makes very little difference.

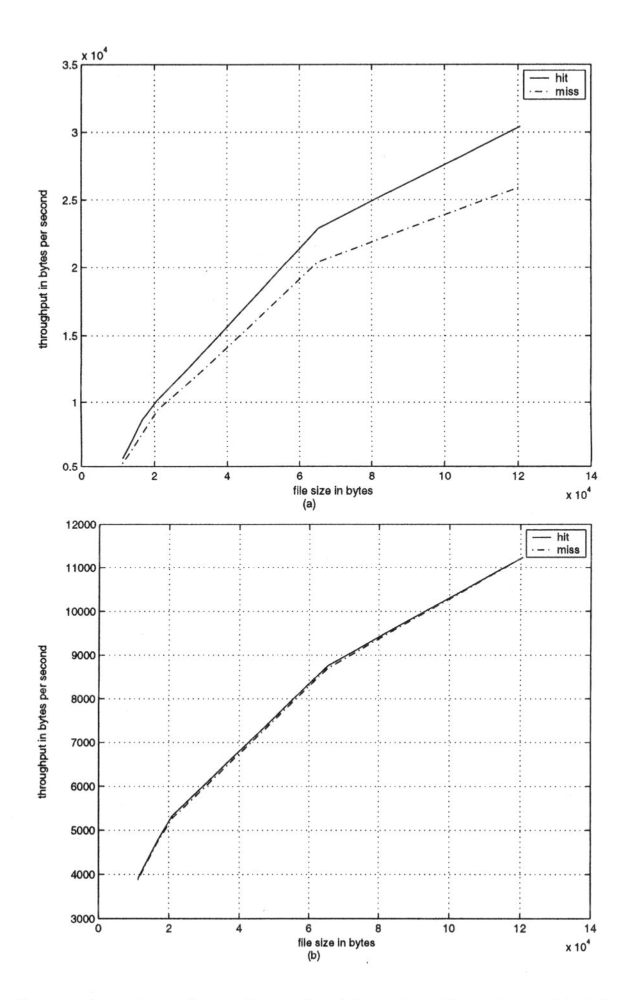


*Figure 8.10.* Comparing throughput for cache hit and cache miss with (a) splitting proxy enabled and (b) splitting proxy disabled

The reason lies in the connection asymmetry we discussed in Section 3. Consider the case where the connection is not split by the proxy. Assuming there is a cache miss (Figure 8.8(c)), since the cache-client connection is much slower (as a result of higher propagation delay) than the server-cache connection, by the time the cache-client connection gets to the first few windows, the entire file could be available in the cache (i.e. the server-cache connection is completed). As an example consider 1 Kbyte packet size and a 7 Kbyte file. Assume that the round trip time of the server-cache connection is 50 msec and the round trip time of the cache-client connection to be 350 msec. As the first packet arrives at the cache it is immediately sent to the client. It takes 2 more round trip times for the whole file to be available in the cache (100 msec), but by this time the first packet has not even reached the client yet. By the time the first acknowledgment gets to the cache from client, the file will be completely available in the cache, so a cache hit and a cache miss have about the same latency and throughput.

As shown in Section 3 when there is asymmetry between the two split segments, the slower segment (as a result of higher propagation delay, smaller initial window size, or higher losses) will dominate the end-to-end performance. In this case the cache-client segment is much slower than the server-cache segment, and clearly dominates the end-to-end performance. Having the file locally in the cache has the effect of "speeding up" the server-cache connection, i.e. this connection is completely eliminated. However since the overall performance is governed by the cache-client connection, whether the server-cache connection is a bit faster or not does not matter much, as shown in Figure 5(b).

Now consider the case where the connection splitting proxy is enabled. Splitting the connection at the gateway results in either three or two segments of an end-to-end connection (Figures 8.8(a) and 8.8(b), respectively). As we have just discussed, if the proxy only splits the connection, then the server-cache connection and the cache-proxy connection would still be much faster than the satellite link and therefore the proxy-client connection would again dominate the overall performance. However, in addition to splitting the connection, the proxy also opens up the window size over the satellite link much faster by using an initial window size of 64 Kbytes and thus bypassing the slow-start stage of normal TCP window evolution. This means that the satellite link is now comparable to or even faster than the server-cache and cache-proxy connections in terms of throughput. For instance, for a file smaller than 64 Kbytes, the entire file could fit into the very first window. Therefore the transfer of the file is constrained by how fast the proxy receives rather than how fast the proxy can send since the window size would be perceived as "unlimited" for such a file. Thus having the file in the cache (much closer to the proxy) would enable the proxy to receive much faster than having to fetch the file remotely, which results in higher throughput and lower latency. This result highlights the importance of optimizing different segments of a split connection. More importantly, such optimization has to be done in a way to reduce asymmetry between the segments, e.g., to bring the slower link faster, which in this case corresponds to using a large initial window size over the satellite link.

## 5.2 Effect of congestion and packet losses

By examining the traces of each file download, we can determine the exact number of losses and retransmissions occurred per connection. However, such losses could involve both random and congestion losses, the distinction of which not directly available to us by only taking measurements at the end point. On the other hand, congestion and losses are highly correlated with increased end-to-end delay, which is observable. In this section we illustrate the relationship between increased file transfer delay and the gain from using connection splitting. In doing so we attempt to understand the relationship between the splitting gain and congestion/losses.

First we repeatedly download a file directly from the server for 2 hours so that the resulting trace may reflect a reasonable range of work load and congestion conditions in the network. We then sort the latency trace of the proxy-enabled connection in descending order, and reorder the proxy-disabled trace (taken at the same time) accordingly. These two cases correspond to that illustrated in Figures 8.8(a) and 8.8(c). Figure 8.11 shows the reordered traces for a 11 Kbyte file and a 120 Kbyte file. Figure 8.12 shows the GoS for these two files. It can be seen that the gain decreases as the latency of the proxy-enabled connection increases. This decrease is much steeper for the 11 Kbyte file and there is a sizable portion of samples showing the proxy-enabled connections underperforming the proxy-disabled connections. This however is not observed in the case of the 120 Kbyte file. In the 120 Kbyte file transfer, split connection always results in smaller latency (higher throughput) throughout the duration of our measurement. We then repeat the same experiment with the same files, but this time files are directly from the cache (corresponding to Figures 8.8(b) and 8.8(d)). Figures 8.13 and 8.14 show the latency and GoS in this case.

There are two main observations from these results. (1) The splitting gain decreases as the file transfer latency increases due to higher loss and/or congestion; (2) Whether the file is directly from the cache or from the remote server, the proxy-enabled connections experience worse performance (higher latency) than the proxy-disabled connections for small file transfers (e.g. 11 Kbyte), for a small portion of the samples. This portion is reduced in the case of a cache hit. The same phenomenon was not observed for large file transfers (e.g., 120 Kbyte). The percentages of negative GoS for different files are listed in Table 8.2. If we define the *probability of performance improvement* as the ratio between the number of samples with a positive GoS and the total number of

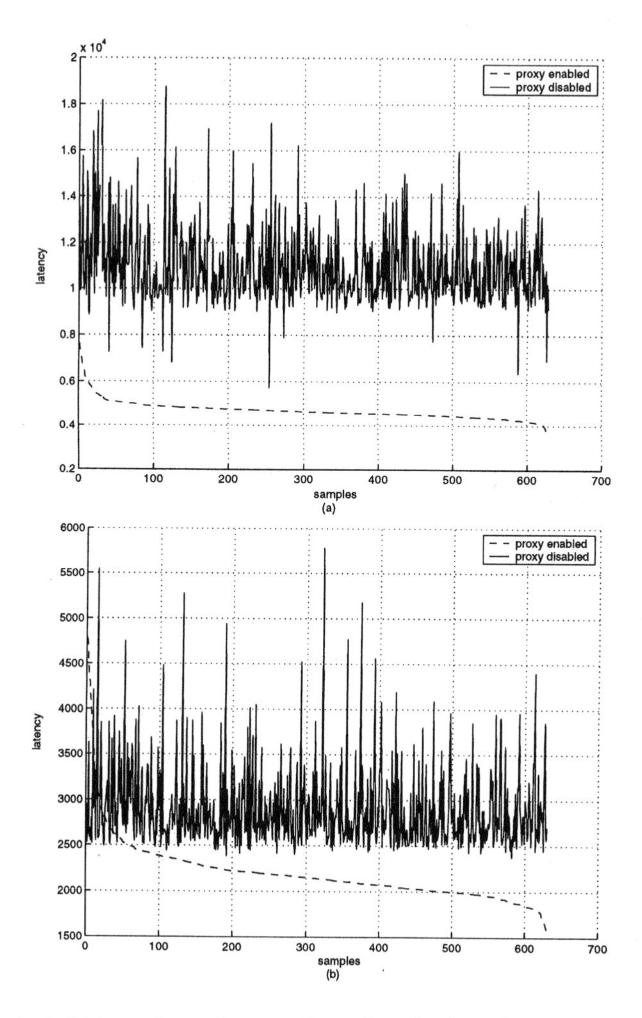


*Figure 8.11.* Sorted latency traces in case of a cache miss for a file of size (a) 120K and (b) 11K

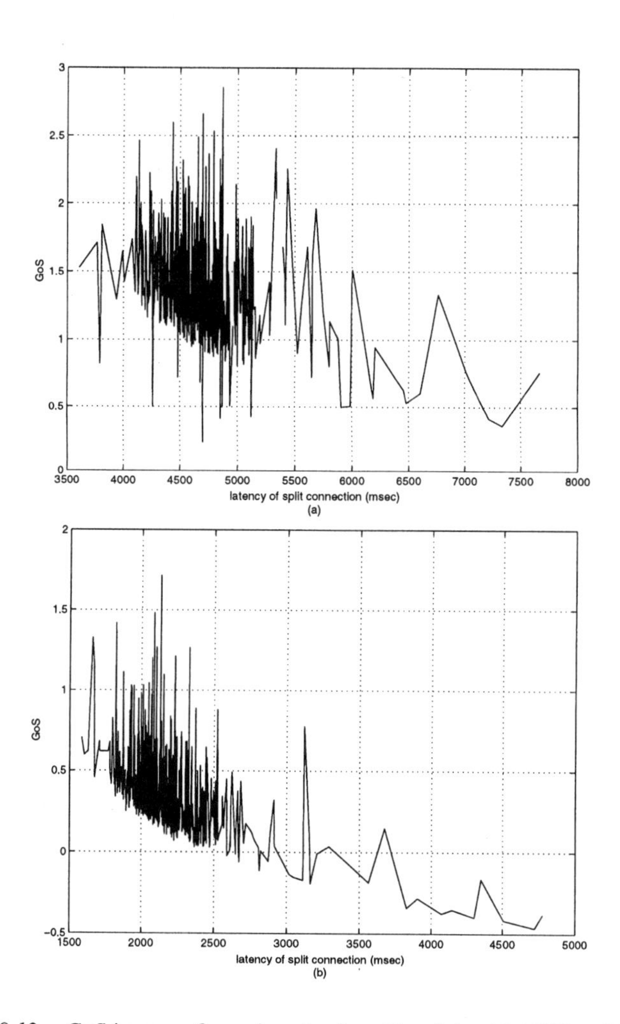


*Figure 8.12.* GoS in case of a cache miss for a file of size (a) 120K and (b) 11K

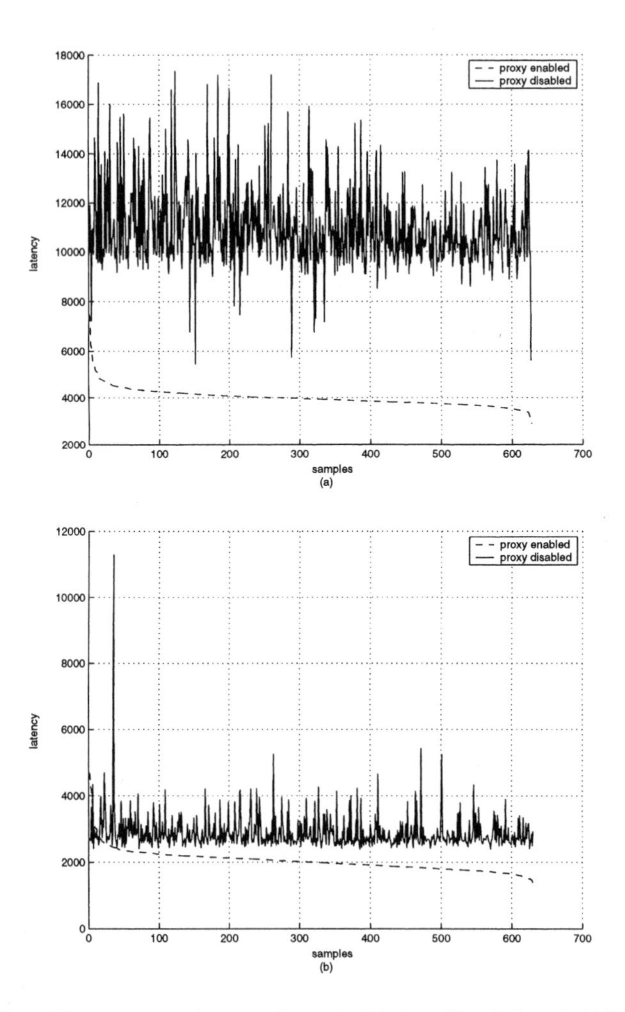


*Figure 8.13.* Sorted latency traces in case of a cache hit for a file of size (a) 120K and (b) 11K

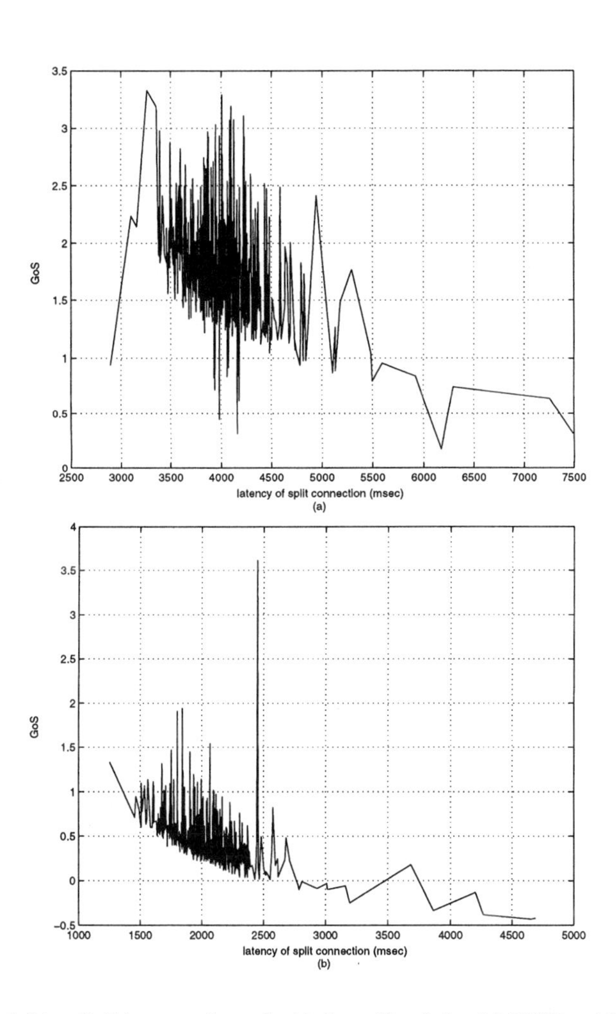


*Figure 8.14.* GoS in case of a cache hit for a file of size (a) 120K and (b) 11K

*Table 8.2.* Percentage of samples where disabling the proxy outperforms enabling the proxy

| *File size (bytes)* | *% of GoS < 0 (cache hit)* | *% of GoS < 0 (cache miss)* |
|---|---|---|
| 11033 | 2.2 | 3.49 |
| 16565 | 0.79 | 1.26 |
| 21027 | 0.0 | 0.48 |
| 65331 | 0.16 | 0.0 |
| 120579 | 0.0 | 0.0 |

samples, then this probability is higher with a larger file transfer (this probability for a given file size is one minus the corresponding term in the table). Note that the average performance of using the splitting proxy is still above that of disabling the proxy.

Since our connections go through a dedicated proxy, the fluctuation in file transfer latency as seen from these traces is mainly due to the fluctuation in work load, congestion and loss situations elsewhere along the path of the connection, i.e., from the server to the cache (in the case of a cache miss), from the cache to the proxy, and from the proxy to the end host [5]. The reason that connection splitting can result in higher latency for small files lies within the relationship between the reduction in latency due to splitting and the increase in latency due to packet losses, both as functions of file size. The reduction in latency by using the proxy is largely due to the fact that the end-to-end TCP is broken down into shorter feedback loops and that a large initial window size is used over the satellite. When the file is small, each packet loss and the time at which it occurs can affect the latency significantly. This effect may not be compensated by the benefit from using the splitting proxy when the file is relatively small since the transfer completes soon afterwords. However, as the file size increases, the benefit of splitting becomes more prominent. In other words, it would take increasingly more packet losses for a split connection to perform at a similar level as a non-split connection. But the probability of this happening decreases as the file size increases since a long connection tends to reflect more average loss rate. This is why we did not observe any such sample over our measurement period. When the file is located in the cache, one segment (server-cache) is eliminated from the path, thus reducing the probability of packet losses and consequently reducing the probability that a particular split connection experiences longer latency than a non-split connection due to different loss occurrences.

In summary, when the file size is large split connections can sustain more losses than non-split connections and still yield shorter latency, thus provide higher probability of performance improvement. When the file is small, the

*Table 8.3.* Throughput of files with different number of embedded objects (in the case of a cache miss)

| *No. of embedded objects* | *Proxy-enabled (bytes/sec)* | *Proxy-disabled (bytes/sec)* | *GoS* |
|---|---|---|---|
| 0 | 32124 | 12544 | 1.56 |
| 7 | 6027 | 3663 | 0.64 |
| 19 | 2701 | 2026 | 0.33 |

split connection is more affected by packet losses and therefore the probability of performance improvement is lower.

## 5.3 Effect of embedded objects and persistent connection

So far all our measurements are taken from pure text files transferred using HTTP. In this section we examine the effect of embedded objects in a file/page, and the effect of using persistent connection.

We first compare the latency for files with the same total transfer size but different numbers of embedded objects. We repeatedly download a text file and two files containing 7 and 19 equal-sized embedded objects respectively over a two-hour period. All three files have a total file size of 65 Kbytes. In downloading these files, HTTP/1.0 [Berners-Lee et al., 1995] without persistent connection is used between the end hosts and the proxy or the cache. No parallel connections are used. The throughput of proxy-enabled and proxy-disabled transfers is shown for both the cache miss (corresponds to Figures 8.8(a) and 8.8(c)) and cache hit (corresponds to Figures 8.8(b) and 8.8(d)) cases, in Tables 8.3 and 8.4, respectively.

We see that when a file contains a larger number of embedded objects the GoS decreases. This result is expected considering our observations that the gain from using the splitting proxy decreases as the file size decreases (before the satellite channel is fully utilized). This is because the time spent in handshake has a bigger portion in the overall latency for smaller files. Thus if we break a large file into many small objects and open a new connection for each of these objects, we expect to see a lower performance gain [6].

We next explore the performance comparison between using HTTP/1.0 and HTTP/1.1 [Fielding et al., 1997]. Web browsers are typically configured to use HTTP/1.0 in the proxy mode where a separate connection is established for each embedded object on a page. HTTP/1.0 generally results in large delay due to the time spent in hand-shaking and low throughput in the slow-start phase of TCP. HTTP/1.1 on the other hand opens up a *persistent* connection which is used to deliver both the base page and subsequent embedded objects. Latency is thus reduced since with a single connection for everything there is

*Table 8.4.* Throughput of files with different number of embedded objects (in the case of a cache hit)

| *No. of embedded objects* | *Proxy-enabled (bytes/sec)* | *Proxy-disabled (bytes/sec)* | *GoS* |
|---|---|---|---|
| 0 | 31451 | 15899 | 0.98 |
| 7 | 6006 | 3845 | 0.56 |
| 19 | 2772 | 2110 | 0.31 |

only one handshake procedure and one slow-start stage of TCP. There has been extensive studies on the performance of different versions of HTTP, see for example [Heidemann et al., 1997; Nielsen et al., 1997].

Here we compare HTTP/1.0 vs. HTTP/1.1 with the option of enabling and disabling the splitting proxy. The connection setup of this part of our measurements corresponds to Figures 8.8(e) and 8.8(f), i.e., the connections do not go through the cache and that the connection is either end-to-end (proxy-disabled) or split into two (proxy-enabled). The resulting GoS for three transfer sizes (base page plus embedded objects) vs. different number of embedded objects are shown in Fig. 8.15. This figure corresponds to 16, 65 and 180 Kbyte file transfer sizes.

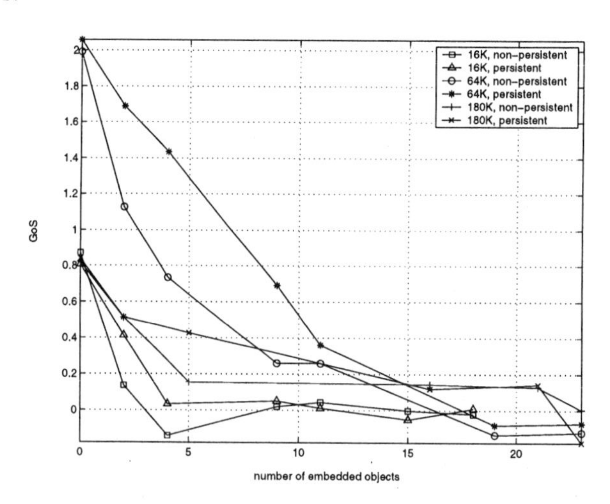


*Figure 8.15.* GoS with varying number of embedded objects.

From these results we see that overall the gain from using connection splitting decreases with increased number of embedded objects in a file (of the same size). This is consistent with previous results in this section. However when persistent connection is used, GoS decreases much faster than the case for non-persistent connections as shown in Fig. 8.15 (by ignoring the first point with 0 embedded object since that is a text file). As the number of embedded objects increases, the splitting gain under persistent connection and non-persistent connection start to converge. The decreasing gain is due to the fact that the client requests each object individually (although within the same connection) after receiving the base page. As the number of objects increases each object becomes smaller, and the overhead in requesting these objects increase. This leads to smaller GoS. Here pipelining is not used. Note that although we did not provide any measurement for the case of persistent connection with pipelining, its effect can be predicted. When pipelining is used, all the requests for embedded objects are sent at the same time after receiving the base page. This is similar to having one large embedded object instead of many small ones. Using Fig. 8.15 (moving toward the left) we expect pipelining to result in higher GoS.

To examine these results from a different angle, let us take a took at the effect of persistent connection in the following way. We define the *Gain of Persistent Connection* (GoP) as follows:

$$\text{GoP} = \frac{\text{Throughput}_{\text{persistent}} - \text{Throughput}_{\text{non-persistent}}}{\text{Throughput}_{\text{non-persistent}}}.$$

Figures 8.16 and 8.17 show the GoP for two transfer sizes, 16 Kbytes and 65 Kbytes respectively.

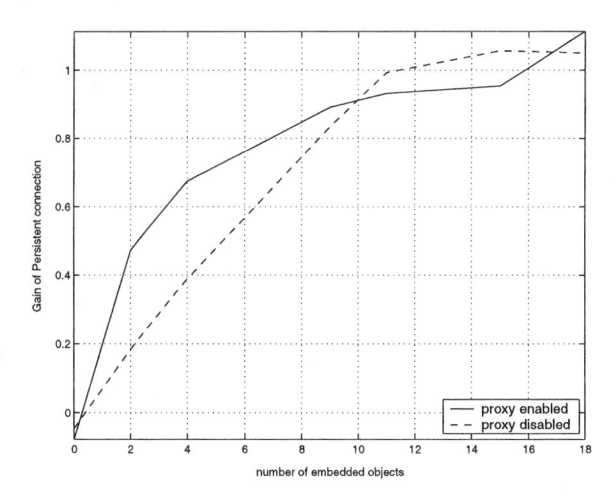


*Figure 8.16.* GoP with total transfer size of 16 Kbytes.

The general observation from this set of figures is that using persistent connection provides higher performance gain over non-persistent connection when the proxy is enabled. In other words, splitting enhances the value of persistent connection. This can be explained as follows. If we denote by $d_p$ and $d_n$ the total transfer delay when persistent connection is used and non-persistent

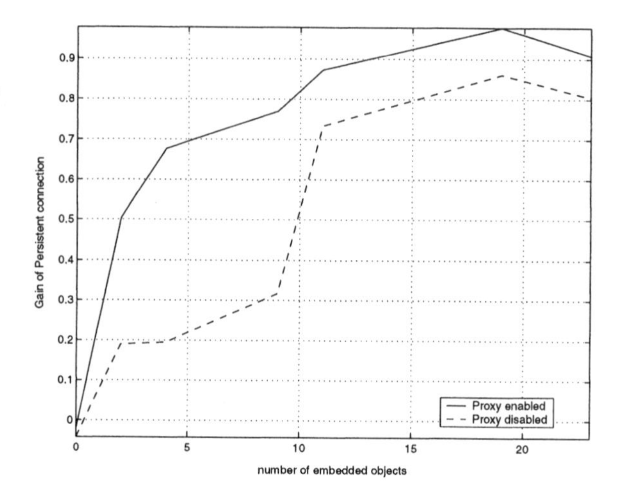


*Figure 8.17.* GoS with total transfer size of 65 Kbytes.

connection is used with the proxy enabled, then approximately

$$d_p = d_n - (j - 1)RTT$$

where $j$ is the number of objects, and RTT is the round trip time between the proxy and the client, denoted by $R_2$. This is because by using persistent connection we save one RTT per object on connection establishment (without pipelining). This approximation also ignores the connection between the server and the proxy. Therefore we have

$$GoP = \frac{\frac{1}{d_p} - \frac{1}{d_n}}{\frac{1}{d_n}} = \frac{(j-1)R_2}{d_p}.$$

Similarly for the proxy disabled case, we have

$$GoP' = \frac{(j-1)(R_1 + R_2)}{d_p'}$$

where $R_1$ is the round trip time between the server and the proxy and $d_p'$ is the total transfer latency when using persistent connection with the proxy disabled. Since $d_p'$ is much larger than $d_p$ for $j$ not too big, we have $GoP > GoP'$. As $j$ increases however, this relationship can reverse. Note that the above analysis is much simplified for the purpose of illustrating the relationship only and not intended to be rigorous.

Finally we compare non-persistent connection with proxy vs. persistent connection without proxy for a total file transfer size of 65 Kbytes, as shown in

*Table 8.5.* Throughput comparison between splitting proxy enabled with persistent connection and disabled with non-persistent connection

| *No. of embedded objects* | *0* | *2* | *4* | *9* | *11* | *19* | *23* |
|---|---|---|---|---|---|---|---|
| Kbyte/sec for persistent, proxy disabled | 6.99 | 5.95 | 4.23 | 3.95 | 4.09 | 3.95 | 3.44 |
| Kbyte/sec for non-persistent, proxy enabled | 21.84 | 10.63 | 6.14 | 3.77 | 2.97 | 1.83 | 1.67 |
| GoS | 2.03 | 0.94 | 0.54 | -0.06 | -0.47 | -0.99 | -0.93 |

Table 8.5. (we did not compare HTTP/1.0 when proxy is disabled vs. HTTP/1.1 when proxy is enabled, as we do not see particular interest in this case.)

We see that the gain from using connection splitting quickly becomes negative, i.e. split connection results in smaller throughput than end-to-end connection. This shows that using a persistent connection is far more important than connection splitting for a file that contains multiple embedded objects (for even a relatively small number) since the time spent in handshake in this case is significant over the satellite link. In other words, the gain from using connection splitting cannot substitute for the performance improvement achieved by having a persistent connection. Therefore in an Internet over satellite system, configuring the browser to run the default HTTP/1.0 with a connection splitting proxy would defeat the purpose of having a proxy, unless all the files are pure text. It is thus crucial that a persistent connection be established between the client and the proxy in such a system.

## 6. IMPLICATIONS ON SYSTEM DESIGN

Throughout the model-based analysis and the measurement-based study we have illustrated various factors that affect the performance gain of using TCP splitting. In this section we highlight the implications of these results on system design options.

We have seen that the performance gain of TCP splitting is highly sensitive to packet losses and diminishes when the proxy becomes congested. In systems that use a splitting proxy, the proxy is usually positioned at a point where all connections have to go through, e.g., located at the NOC in the satellite system we show. Therefore it is very important to properly provision such a system to minimize the probability of buffer overflow and queuing delay at the proxy. In particular, in a satellite system the server typically sends much faster than the proxy. Thus some form of dynamic flow control at the proxy is critical in regulating the server sending rate, e.g., by advertising a smaller receive window size.

TCP splitting separates the end-to-end connection into different segments. However, these segments are not completely uncoupled. The slowest segment will constrain the other segments and dominate the end-to-end performance. The proxy cannot forward client data it has not received from the server. Therefore if the proxy is constrained by slow receiving then increasing the rate at which it receives will improve the overall performance. Similarly, if the proxy is constrained by slow sending then we should increase the rate at which it sends. This is why cache hits improve the throughput when the splitting proxy is enabled. On the other hand if we only use the cache and disable the splitting proxy, we will have a cache-client connection highly constrained by slow sending due to the large propagation delay. Thus having a large initial window size, or for that matter any transmission control scheme that would speed up sending over the satellite link is critical. To summarize in designing such systems it is very important to optimize each split segment to reduce the asymmetry among them so that one is not constrained by the other.

## 7. CONCLUSION

In this chapter we examined using TCP connection splitting/spoofing as a way of improving TCP performance in a heterogeneous environment and more specifically over the satellite. This scheme breaks an end-to-end connection into two segments so that each segment can be optimized separately. We presented a model-based analysis to illustrate the TCP splitting dynamics and the performance gain under various conditions. We then presented a measurement-based study via a specific example of a satellite system. In our experiments, not all parameters are controllable, e.g., the bit error rate over the satellite link, the amount of traffic and congestion in the network, etc. However, by properly isolating elements that contribute to the result, we were able to confirm our analysis and earlier work on connection splitting, and reveal additional insights. In particular we examined the combined effect of caching and connection splitting and the use of persistent connection. We summarize our observations and conclusions as follows.

In general TCP connection splitting is a valid approach to improve TCP throughput and reduce file transfer latency over the satellite. An end-to-end connection can be compensated by increasing the initial window size. However, it requires significantly larger initial window size even for reasonably small file sizes, which makes it less practical in real applications. For file transfers that do not fill up the satellite link capacity, as the file increases, the gain from using proxy also increases. However, once the pipe capacity is reached, the difference in latency between using end-to-end connection and using proxy stays constant, thus the difference in throughput decreases with the increase in file size.

A proxy achieves the effect of localizing error/loss recovery and improves the throughput and reduces latency of a connection when losses are present. On the other hand, when a connection is broken in two, the slower one always dominates the overall performance, and as this dominance increases, the gain from using the proxy is again reduced. These results imply that while optimization of separate parts of a connection (segregated by the proxy) is important, it is equally important to minimize the "asymmetry" between these parts, especially in a heterogeneous environment.

When the proxy becomes the bottleneck, the gain from using the proxy quickly diminishes. The gain is also very sensitive to congestion and packet losses, especially for small files, while larger files have a much greater probability of performance gain by using TCP splitting. On the one hand having a cache hit will alleviate such sensitivity, on the other hand using connection splitting proxy enhances the benefit of caching by speeding up the transfer over the satellite link.

For transfer of files with embedded objects, the performance gain of using connection splitting decreases as the number of embedded objects increases. Using TCP splitting also results in higher performance gain of persistent connection over non-persistent connection. Lastly, although connection splitting improves throughput, it is no substitute for persistent connection. The best performance is achieved by using both the splitting proxy and persistent connection (HTTP/1.1) between the proxy and the client.

## ACKNOWLEDGMENT

Most of the material in this chapter is taken from the following two papers [Liu and Ehsan, 2002; Ehsan et al., 2002]:

Liu, M. and Ehsan, N. (2002). Modeling TCP performance with proxies. *International Workshop on Wired/Wireless Internet Communications (WWIC), in conjunction with International Conference on Internet Computing (IC'02).*

Ehsan, N., Liu, M., and Ragland, R. (2002). Measurement Based Performance Analysis of Internet over Satellite. *Proc. International Symposium on Performance Evaluation of Computer and Telecommunication Systems (SPECTS 2002).*

## Notes

1. The terms TCP splitting and TCP spoofing are often used interchangeably in the literature. Strictly speaking splitting refers to breaking up a connection and spoofing refers to faking an address. They are often used together in the context of satellite because in splitting a connection a transparent proxy typically spoofs the end points' addresses. Since our interest in this chapter is in the end-to-end performance as a result of segmented connections (either at the transport layer or at the application layer, see later parts of this chapter), we are only concerned with the fact that a connection is split rather than how the addresses are handled. However we will follow the convention here and use both terms interchangeably.

2. In general, shortening the TCP feedback loop does not have to come at the price of breaking the end-to-end semantics, e.g., feedback from the proxy can be a signal for the server to increase its window, rather than to purge the retransmission buffer. This of course may imply that the proxy is not transparent to the end users anymore and involve modifications to the existing systems, which is out of the scope of out study here. In subsequent sections we will ignore whether the proxy spoofs addresses or not since it does not affect our analysis, and instead focus on a general model of server-proxy-client communication. The server and the client can either have an end-to-end TCP connection, where the proxy simply acts as a router, or they can have a connection in which the proxy breaks up the connection in two.

3. For a cache hit on the file request, the content is retrieved directly from the cache. In this case the connection model is simply end-to-end from the client to the proxy, with a fraction of the entire server-client round-trip time.

4. As pointed out in [Lakshman et al., 2000], asymmetry between the data path and ACK path increases TCP's sensitivity to random packet losses, thus adjustments to our models are needed. Our measurement-based study in the later sections is based on an asymmetric satellite link.

5. Note that in general, connection splitting leading to worse performance can be caused by excessive congestion and delay at the proxy, which is only experienced by split traffic, but not by end-to-end traffic. This can happen if the proxy handles many split connections at the transport layer, while the end-to-end traffic simply goes through the IP queue and is unaffected. However, since we use a dedicated proxy the increase in delay and loss incurred by splitting is minimal.

6. It should be noted that the measurements in Tables 8.3 and 8.4 are taken over different time periods. Thus numbers from different tables are not directly comparable.

## References


Allman, M., Dawkins, S., Glover, D., D.Tran, Henderson, T., Heidemann, J., Touch, J., Kruse, H., Ostermann, S., Scott, K., and Semke, J. (2000). Ongoing TCP Research Related to Satellites. *IETF RFC 2760.*

Allman, M. and Paxson, V. (1999). On Estimating End-to-end Network Path Properties. *ACM SIGCOMM.*

Badrinath, B. and Sudame, P. (1996). To Send or Not to Send: Implementing Deferred Transmissions in a Mobile Host. *Proc. IEEE ICDCS*, pages 327–333.

Bakre, A. and Badrinath, B. (1995). I-TCP: Indirect TCP for Mobile Hosts. *Proc. IEEE ICDCS*, pages 136–143.

Bakre, A. V. and Badrinath, B. R. (1997). Implementation and Performance Evaluation of Indirect TCP. *IEEE/ACM Transactions on Computers*, 46(3):260–278.

Balakrishnan, H., Padmanabhan, V. N., and Katz, R. H. (1997a). The Effects of Asymmetry on TCP Performance. *ACM/IEEE International Conference on Mobile Computing and Networking (MobiCom'95).*

Balakrishnan, H., Padmanabhan, V. N., Seshan, S., and Katz, R. H. (1997b). A Comparison of Mechanisms for Improving TCP Performance over Wireless Links. *IEEE/ACM Transactions on Networking*, 5(6):756–769.

Balakrishnan, H., Seshan, S., Amir, E., and Katz, R. H. (1995). Improving TCP/IP Performance Over Wireless Networks. *ACM/IEEE International Conference on Mobile Computing and Networking (MobiCom'95)*, 2(11).

Berners-Lee, T., Fielding, R., and Frystyk, H. (1995). Hypertext Transfer Protocol – HTTP/1.0. *IETF RFC 1945.*

Bharadwaj, V. G. (1999). Improving TCP Performance over High-Bandwidth Geostationary Satellite Links. Technical Report MS 99-12, Institute for Systems Research, University of Maryland, College Park.

Biaz, S. and Vaidya, N. H. (1999). Discriminating Congestion Losses From Wireless Losses Using Inter-arrival Times at the Receiver. *Proc. IEEE ASSET*, pages 10–17.

Cáceres, R. and Iftode, L. (1995). Improving the Performance of Reliable Transport Protocol in Mobile Computing Environment. *IEEE J-SAC*, 13(5):850–857.

Cardwell, N., Savage, S., and Anderson, T. (2000). Modeling TCP Latency. *IEEE INFOCOM*.

Ehsan, N., Liu, M., and Ragland, R. (2002). Measurement Based Performance Analysis of Internet over Satellite. *Proc. International Symposium on Performance Evaluation of Computer and Telecommunication Systems (SPECTS 2002)*.

Fielding, R., Gettys, J., Mogul, J., Frystyc, H., and Berners-Lee (1997). Hypertext Transfer Protocol – HTTP/1.1. *IETF RFC 2068*.

Heidemann, J., Obraczka, K., and Touch, J. (1997). Modeling the Performance of HTTP over Several Transport Protocols. *IEEE/ACM Transactions on Networking*, 5(5).

Ishac, J. and Allman, M. (2001). On the Performance of TCP Spoofing in Satellite Networks. *IEEE MILCOM*.

Kurose, J. and Rose, K. (2000). *Computer Networking, A Top-Down Approach Featuring the Internet*. Addison-Wesley.

Lakshman, T. V. and Madhow, U. (1997). The Performance of TCP/IP for Networks with High Bandwidth-Delay Products and Random Loss. *IEEE/ACM Transactions on Networking*, 5(3):336–350.

Lakshman, T. V., Madhow, U., and Suter, B. (2000). TCP Performance with Random Loss and Bidirectional Congestion. *IEEE/ACM Transactions on Networking*, 8(5):541–555.

Liu, M. and Ehsan, N. (2001). Modeling TCP performance with proxies. *Technical Report, EECS Department, University of Michigan, Ann Arbor*.

Liu, M. and Ehsan, N. (2002). Modeling TCP performance with proxies. *International Workshop on Wired/Wireless Internet Communications (WWIC), in Proc. International Conference on Internet Computing (IC'02)*.

Mathis, M., Mahdavi, J., Floyd, S., and Romanow, A. (1996). TCP Selective Acknowledgement Options. *IETF RFC 2018*.

Nielsen, H. F., Gettys, J., Baird-Smith, A., Prud'hommeaux, E., Lie, H. W., and Lilley, C. (1997). Network Performance Effects of HTTP/1.1, CSS1, and PNG. Technical report, W3C. URL: `http://www.w3.org/Protocols/HTTP/Performance/Pipeline.html`.

Padhye, J., Firoiu, V., Towsley, D. F., and Kurose, J. F. (2000). Modeling TCP Reno Performance: A Simple Model and Its Empirical Validation. *IEEE/ACM Transactions on Networking*, 8(2):133–145.

Parsa, C. and Garcia-Luna-Aceves, J. J. (1999a). Improving TCP Performance Over Wireless Network at The Link Layer. *ACM Mobile Networks & Applications Journal.*

Parsa, C. and Garcia-Luna-Aceves, J. J. (1999b). TULIP: A Link-Level Protocol for Improving TCP over Wireless Links. *Proc. IEEE WCNC'99*, pages 1253–1257.

Ratnam, K. and Matta, I. (1998a). Effect of Local Retransmission at Wireless Access Points on The Round Trip Time Estimation of TCP. *Proc. 31st Annual Simulation Symp.*, pages 150–156.

Ratnam, K. and Matta, I. (1998b). WTCP: An Efficient Mechanism for Improving TCP Performance Over Wireless Links. *Proc. IEEE ISCC*, pages 74–78.

Rodriguez, P., Sibal, S., and Spatscheck, O. (2000). TPOT: Translucent Proxying of TCP. Technical report, AT & T labs-Research and EURECOM Technical Report.

Vaidya, N. H., Mehta, M., Perkins, C., and Montenegro, G. (1999). Delayed Duplicated Acknowledgments: A TCP-Unware Approach to Improve Performance of TCP over Wireless. *Technical Report 99-003, TAMU.*

# Chapter 9

# SCHEDULING DATA BROADCAST

Shu Jiang
*Dept. of Computer Science, Texas A&M University*

Nitin H. Vaidya
*Dept. of Electrical and Computer Engineering, University of Illinois at Urbana-Champaign*

**Abstract** Data broadcast is an effective method of information dissemination in satellite networks. Communications in satellite networks are usually asymmetric – the *downstream* communication capacity, from information server to clients, is much greater than the *upstream* communication capacity, from clients to server. It is not efficient (and sometimes not possible) for clients to send explicit requests to the server and for the server to serve requests individually. In data broadcast approach, the server broadcasts data items periodically over a broadcast channel that all clients listen to, serving a large number of clients demanding for same information simultaneously. One challenge in implementing this solution is to determine the broadcast schedule, such that the clients receive the best quality of service.

In this chapter, we study the optimal broadcast scheduling problem in a *pure push* data broadcast system. Our study is different from other researchers in that we assume both persistent user model and impatient user model. For each model, we define different performance metrics to address the most important user concerns. The properties of schedules that optimize each metric are derived through theoretical analysis. Based on theoretical results, we propose a heuristic algorithm for producing near-optimal schedules on-line. Performance evaluation results are also presented.

**Keywords:** data broadcast schedule, on-line scheduling algorithm, performance evaluation

## 1. INTRODUCTION

One important application of satellite network technologies is to disseminate information to a large population of users over a large geographical area. Mechanisms for efficient data dissemination are of significant interest. There are many options of such mechanisms among which data broadcast is one of the most direct, low-overhead and easy-to-implement solutions [Acharya et al., 1995; Ammar and Wong, 1985; Ammar and Wong, 1987; Gescei, 1983; Herman et al., 1987; Zdonik et al., 1994; Imielinski et al., 1994; Imielinski et al., 1996; Banerjee and Lee, 1994]. Data broadcasting is especially suitable for transmitting information of common interest to a group of users.

In this chapter, we address some scheduling issues in a data broadcast system. For the purpose of our study, we assume a simple system model in which a database server transmits data items over a broadcast channel shared by a large number of users. All users who have data requests are tuned to the same channel waiting for the moment when the requested items are transmitted. When an item is transmitted on the broadcast channel, all pending requests for the item are served. Since users have diverse requests and the broadcasts of data items are multiplexed over the channel, the server should make the broadcast schedule properly to optimize the system performance. Typically, the following two metrics are used to evaluate broadcast schedules and scheduling algorithms:

- Access time: This is the amount of time elapsed from when a user generates a data request until when the request is served. It is important to minimize the access time so as to decrease the idle time at the user. However, the performance goal can only be pursued in a statistical sense. Other researchers have considered the problem of minimizing the mean access time over all requests. In our study, we also address the problem of minimizing the variance of access time as well as the trade-off between both.

- Tuning time: This is the amount of time a user must listen to the channel until it receives the information it needs. It is important to minimize the tuning time, because it directly affects the power consumption of a wireless device. The tuning time is usually equal to the access time unless (1) the user knows the broadcast schedule in prior so that it can put itself in *doze* mode, or (2) the user becomes impatient after waiting for some time and retrieves the data request. In this chapter, we propose an online scheduling approach with which the server determines the broadcast schedule "on the fly". Since there is no prior schedule in place, we do not consider case (1) and interested readers are referred to [Imielinski et al., 1994] for an introduction of indexing technique.

The rest of this chapter is organized as follows. Section 2 introduces the basic system model we use in our theoretical analysis and define performance metrics. Section 3 presents the optimality conditions that a broadcast schedule must satisfy to optimize a particular performance metric. Based on the analytical results, we propose on-line scheduling algorithms in section 4 to implement near-optimal schedules. The performance evaluation results of the scheduling algorithms are presented in section 5. In section 6, we summarize our work and conclude the chapter.

## 2. THE BASIC MODEL

In this chapter, we consider the *pure push* model wherein the broadcast schedule by the server is a function of *request probability distribution* for the data items in the server's database – the request probability distribution provides the server with a measure of the popularity of various data items. Assuming that there are $M$ data items numbered from 1 through $M$, request probability $p_i$ of data item $i$ is the probability that item $i$ is requested in a new request from user, where $\sum_{i=1}^{M} p_i = 1$. The actual requests pending at a given time are not known to the server. However, our results can be easily extended to the *pure pull* model wherein the broadcast schedule is a function of the requests known to be pending at a given time (thus, in the *pure pull* model, the server is aware of all the pending requests). An algorithm for the pure pull model can be obtained by replacing the request probability $p_i$ by number of pending requests for item $i$ in the algorithm presented later in the chapter. In addition, we define length of data item $i$, $l_i$, as the time required to broadcast item $i$.

Similar to some past work on broadcast scheduling, users are assumed to submit independent requests according to a Poisson process with rate $\lambda$ [Ammar and Wong, 1987; Su and Tassiulas, 1997]. Thus, requests for item $i$ are generated according to a Poisson process with rate $\lambda_i = p_i\lambda$. Note that, when different requests are correlated, or when a user may make multiple requests at the same time, the analysis presented below will not apply.

We consider two different user behavior models and define different metrics to evaluate the broadcast schedules respectively.

### 2.1 Persistent User Model

This is the model adopted by most researchers, where once upon a data request is generated, the user listens to the channel until the requested item appears. Therefore, for each request, the access time is equal to the tuning time.

We define the following metrics:

- Mean access time $\mu$: This is the expected value of request access time.

- Variance of access time $\delta^2$: This is the expected variance of request access time.

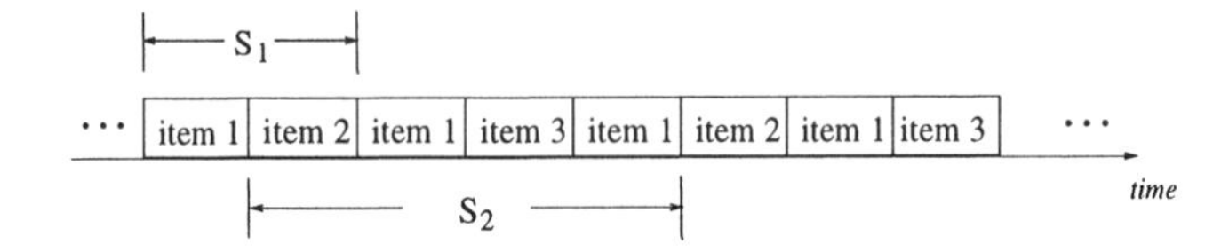


*Figure 9.1.* An example broadcast schedule

## 2.2 Impatient User Model

With impatient users, a pending request has two possible outcomes: satisfied or dropped. To facilitate the analysis, we assume that the length of time that a user will wait before dropping a request, $w$, is an exponentially distributed random variable with mean value $1/\tau$.

We define the following metrics:

- Service ratio $R$: This is the expected fraction of requests that are serviced among all requests that are generated.
- Mean tuning time $W$: This is the expected time that a user will spend on listening to the channel for a request no matter whether the request is satisfied or dropped eventually.

# 3. THEORETICAL RESULTS

For the theoretical development, We consider broadcast schedules with Equal Spacing property. In such a schedule, the transmissions of any particular item on broadcast channel are equally spaced. Let $s_i, 1 \leq i \leq M$ be the spacing between consecutive instances of item $i$. We refer to the vector $< s_1, s_2, \ldots, s_M >$ as *schedule vector*. Figure 9.1 is a snapshot of a broadcast schedule that illustrates the concept of item spacing on broadcast.

Given a schedule vector, the performance metric values can be derived theoretically. In Table 9.1, we represent all metrics as functions of a schedule

*Table 9.1.* Metric and optimality condition

| | *Metric* | *Optimality Condition* |
|---|---|---|
| Persistent user model | $\mu = \frac{1}{2}\sum_{i=1}^{M} s_i p_i$ | $\frac{s_i^2 p_i}{l_i} = constant$ |
| | $\sigma^2 = \frac{1}{3}\sum_{i=1}^{M} p_i s_i^2 - \mu^2$ | $\frac{p_i s_i^2}{l_i}(\frac{2}{3}s_i - \mu) = constant$ |
| Impatient user model | $R = \sum_{i=1}^{M} \frac{p_i}{\tau s_i}(1 - e^{-\tau s_i})$ | $\frac{p_i}{l_i}(\tau s_i e^{-\tau s_i} + e^{-\tau s_i} - 1) = constant$ |
| | $W = \frac{1}{\tau}(1 - R)$ | same as above |

vector and give optimality conditions that a broadcast schedule must satisfy to optimize each metric. Detailed derivations are given in appendixes.

From Table 9.1, the service ratio $R$ is maximized by the same schedule vector that minimizes the mean tuning time $W$. However, there is no such relationship between the mean access time and the variance of access time. Actually, the minimization of $\mu$ and the minimization of $\sigma^2$ are two contradictory goals. In later sections, we explore the trade-off between the two objectives.

Even if we have revealed the optimality condition for each metric, it is hard to obtain a closed-form solution for the optimal schedule vector (under the constraint that the $M$ items use the available broadcast bandwidth). Therefore, we propose a heuristic-based on-line scheduling algorithm in the next section.

## 4. ON-LINE SCHEDULING ALGORITHM

Whenever the server is ready to transmit a new item, it calls the on-line algorithm presented here. The on-line algorithm determines the item to be transmitted next using a decision rule - this decision rule is motivated by an optimality condition depending on which metric is to be optimized. As an example, we show how the scheduling algorithm works when we try to minimize the mean access time.

Let $Q$ be the current time and $R_i$ be the time when item $i$ was most recently transmitted. $R_i$ is initialized to -1 for each item $i$ and updated every time when item $i$ is transmitted. Define $F_i$ as

$$F_i = (Q - R_i)^2 \frac{p_i}{l_i} \tag{9.1}$$

The above definition of $F_i$ is adapted from the optimality condition for $\mu$: $\frac{s_i^2 p_i}{l_i} = constant$. According to the optimality condition, the produced schedule should keep the values of all $F_i$'s as close to each other as possible if not the same. Therefore, the server picks the item with maximum $F$-value to broadcast as follows.

**Algorithm 1** Algorithm for reducing mean access time

1: For each item $i$, $1 \leq i \leq M$, update the value of $F_i$.
2: Determine maximum $F_i$ over all items. Let $F_{max}$ denote the maximum value.
3: Choose item $j$ such that $F_j = F_{max}$. If this equality holds for more than one item, choose any one of them arbitrarily.
4: Broadcast item $j$.
5: $R_j = Q$.

By changing the definition of $F_i$, this scheduling algorithm can produce schedules to achieve other optimization goals. For example, to minimize the variance of access time, we can use the following definition:

$$F_i = \frac{p_i(Q - R_i)^2}{l_i}\Big(\frac{2}{3}(Q - R_i) - \frac{1}{2}\sum_{i=1}^{M} p_i(Q - R_i)\Big) \tag{9.2}$$

In general, minimal mean and minimal variance of access time are unlikely to be achieved by same schedule. Therefore, we propose a new definition of $F_i$ with the objective of achieving a trade-off between a small mean and a small variance of access time, which is

$$F_i = (Q - R_i)^\alpha \frac{p_i}{l_i} \tag{9.3}$$

The above definition is motivated by observations we make on the optimality conditions for mean and for variance of access time. When $\alpha = 2$, definition 9.3 becomes definition 9.1 for minimizing the mean access time. When $\alpha = 3$, the dominant exponent of $(Q - R_i)$ in definition 9.3 is same as that in definition 9.2, which is 3. Therefore, we expect that the produced schedule will have performance approaching the schedule produced with definition 9.2, which is aimed at minimizing the variance of access time. By varying $\alpha$, we hope to find a schedule with both small mean and small variance of access time.

Assuming impatient user model, we can use the following definition to maximize the service ratio and minimize the mean tuning time:

$$F_i = \frac{p_i}{l_i}(\tau(Q - R_i)e^{-\tau(Q-R_i)} + e^{-\tau(Q-R_i)} - 1) \tag{9.4}$$

By using different definitions of $F_i$, we obtain a group of scheduling algorithms that pursue different performance goals. In the next section, we evaluate the performance of these scheduling algorithms by simulation.

## 5. PERFORMANCE EVALUATION

We simulate a data broadcast system in which the server uses the on-line algorithms presented above to schedule broadcasts. We also simulate the user request generation process and measure the performance metrics. In each simulation run, at least 1 million requests are served.

In our simulations, the demand probabilities follow Zipf distribution, with item 1 being the most frequently requested, and item $M$ being the least frequently requested. The Zipf distribution may be expressed as follows:

$$p_i = c\left(\frac{1}{i}\right)^\theta, 1 \leq i \leq M$$

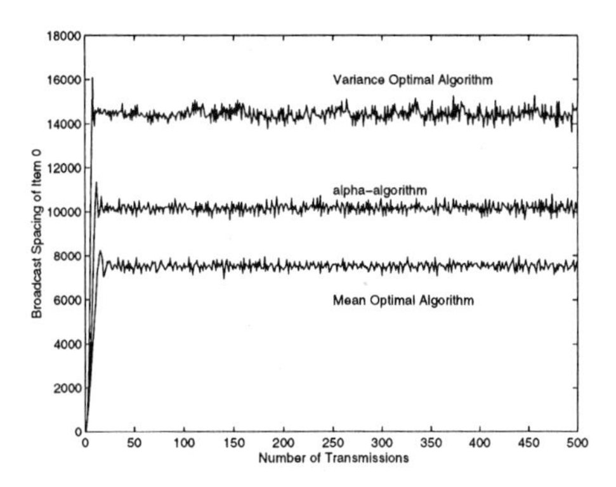


*Figure 9.2.* The broadcast spacing of item 1

where $c = \frac{1}{\sum_{i=1}^{M}(\frac{1}{i})^{\theta}}$ is a normalizing factor, and $\theta$ is a parameter named *access skew coefficient*. When $\theta = 0$, Zipf distribution reduces to a uniform distribution with each item equally likely to be requested. However, the distribution becomes increasingly "skewed" as $\theta$ increases (that is, the difference among items with respect to the degree of popularity becomes more significant).

In our simulations, we consider three length distributions:

- Equal Length Distribution:

$$l_i = 1$$

- Increasing Length Distribution:

$$l_i = l_{min} + \frac{(i-1)(l_{max} - l_{min})}{M-1}$$

with $l_{min} = 1$ and $l_{max} = 250$.

- Decreasing Length Distribution:

$$l_i = l_{max} - \frac{(i-1)(l_{max} - l_{min})}{M-1}$$

with $l_{min} = 1$ and $l_{max} = 250$.

However, due to space limit, we only present simulation results for the equal length distribution setting in this chapter. In fact, based on our results, the system performance is not sensitive to the item length distribution.

## 5.1 Validation of algorithm

Since all our theoretical analysis is based on an assumption that the broadcast schedule has equal spacing property, we first verify that the proposed heuristic-based scheduling algorithm produces such a schedule indeed. Figure 9.2 plots

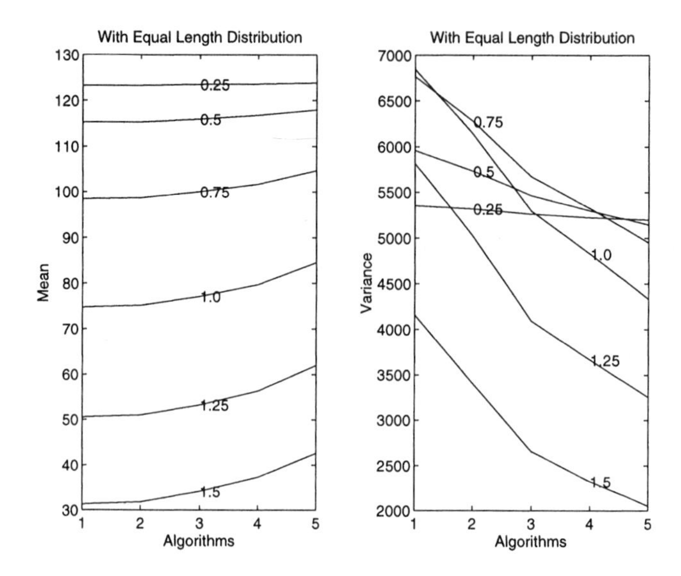


*Figure 9.3.* Performance of different algorithms (1:Mean Optimal algorithm, 2:$\alpha$ algorithm with $\alpha = 2.2$, 3:$\alpha$ algorithm with $\alpha = 2.6$, 4:$\alpha$ algorithm with $\alpha = 3$, 5:Variance Optimal algorithm)

the first 500 transmission intervals of one item that we recorded in three simulation runs. In each run, the particular item is broadcast at a constant interval approximately.

## 5.2 Persistent user case

In this set of experiments, we evaluate the performance of scheduling algorithms when all users are persistent users. We examine three scheduling algorithms: the Mean Optimal algorithm, the Variance Optimal algorithm and the $\alpha$ algorithm. The difference between them is on the use of $F_i$ definition. The Mean Optimal algorithm uses definition 9.1 and tries to minimize the mean access time. The Variance Optimal algorithm uses definition 9.2 and tries to minimize the variance of access time. The $\alpha$ algorithm uses definition 9.3 and tries to achieve a trade-off between the two optimization goals.

Figure 9.3 plots the measured mean and variance of access time in different simulation runs when different algorithms are used to schedule the broadcast. In the figure, the number marked on each curve is the $\theta$ value. In all simulation runs, $M = 250$.

In general, all algorithms position themselves as expected. In all cases, the lowest mean access time is experienced when server uses the Mean Optimal algorithm, and the lowest variance of access time when Variance Optimal algorithm. $\alpha$-algorithms lies between these two algorithms, with higher mean and

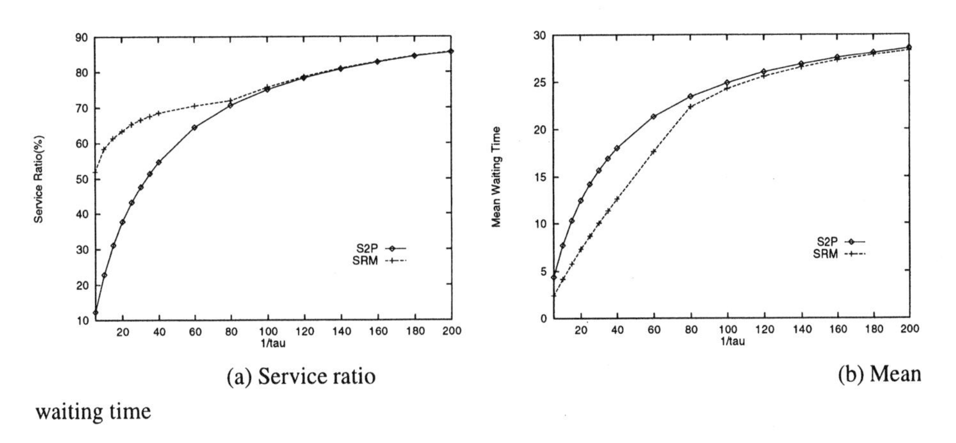


(a) Service ratio (b) Mean waiting time

*Figure 9.4.* System performance as request adjourn time varies

lower variance of access time. When $\alpha$ changes from 2.2 to 2.6 and then to 3, the measured mean access time increases while the variance drops.

When $\theta = 0.25$, the user requests are just slightly skewed. In this case, all algorithms show similar performance and the measured mean access time is very close to the theoretically minimal value. When $\theta$ is between 0.25 and 1.0, the user request skewness is moderate. For example, when $\theta = 0.75$ and $M = 250$, about half user requests are for top 33 items and the other half requests go to the remaining 217 items. In this case, the Optimal Variance algorithm shows better performance than other algorithms with a slightly higher mean access time and a significantly lower variance. When $\theta$ is larger than 1.0, the request skewness becomes serious. For example, when $\theta = 1.5$ and $M = 250$, top 33 items are demanded by 91% user requests. In this case, the $\alpha$ algorithm produces ideal schedules that show low mean as well as low variance of access time.

## 5.3 Impatient user case

In this set of experiments, we evaluate the system performance when users are not persistent and intolerant of long waiting. We evaluate the performance of the Service Ratio Minimized algorithm which uses definition 9.4 of $F_i$. For comparison purpose, we also evaluate the performance of the Mean Optimal algorithm which does not take user impatience into account. In the following figures, *S2P* stands for the Mean Optimal algorithm and *SRM* stands for the Service Ratio Minimized algorithm.

Figure 9.4 plots measured service ratio and mean tuning time values against varied $1/\tau$ values (when $M = 100$ and $\theta = 1$). In Figure 1(a), when $1/\tau$ is large enough, the service ratio is higher than $801/\tau \rightarrow \infty$, the service ratio will be 100when $1/\tau$ is small meaning that users cannot afford to wait long,

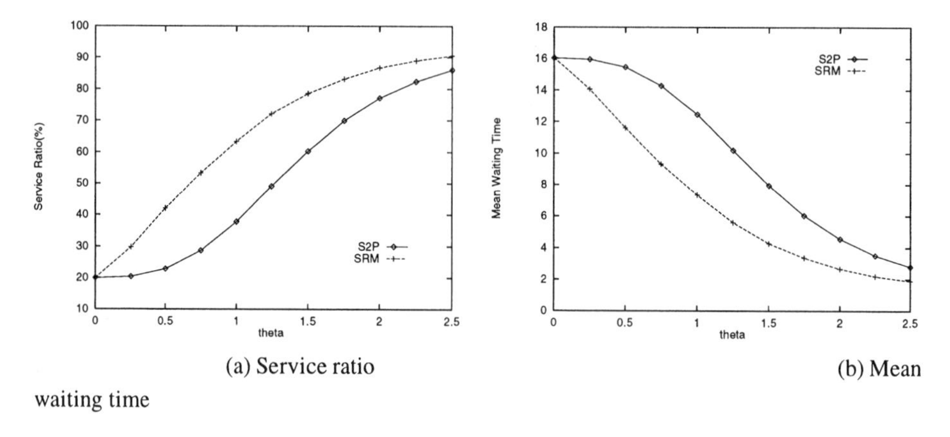


(a) Service ratio (b) Mean waiting time

*Figure 9.5.* System performance as request skewness varies

the "minimal mean access time" schedule causes less than 30served while in the same situation, the *SRM* algorithm produces much higher service ratio.

Next we examine how skewness in the data request pattern would affect the performance of scheduling algorithms. In Figure 9.5, we plot the measured service ratio and mean tuning time as functions of $\theta$ respectively (when $M = 100$ and $1/\tau = 20$). When $\theta = 0$, the user requests are uniformly distributed on the data item set. It is not surprise that both schedules do not serve user requests well, indicated by the lowest service ratio and the highest mean tuning time. In fact, when all data items are equally possible to be requested, the best schedule is the so-called "flat schedule" where all data items are broadcast in a round-robin fashion. When request skewness is moderate, the *SRM* algorithm generates better schedule than the *S2P* algorithm in terms of the service ratio and the mean tuning time. But when request skewness is serious, the two algorithms yield similar performance. The reason is that when most of user requests are targeted to a few "hot" items, both algorithms schedule the transmission of these items more frequently so that most of user requests are satisfied before users lose their patience. This is indicated by the high service ratio.

## 6. CONCLUSIONS

In this chapter, we address the scheduling problem in a data broadcast system. Unlike other studies, we consider both persistent user model and impatient user model. When all users are persistent and keep waiting until requests are served, we define mean access time and variance of access time as metrics for evaluating the performance of broadcast schedules. Based on the definitions, We derive the respective properties that the optimal schedule must possess in order to minimize each metric. When users are not persistent and may leave after waiting for some time if the requested items are still not broadcast, we define service ratio and mean tuning time as schedule performance metrics. Similarly, we derive the optimal schedule property for maximizing the service ratio and minimizing the mean tuning time. Based on our analytical results, we propose a heuristic-based on-line scheduling algorithm to implement the optimal schedules. Extensive simulation is conducted and the results are encouraging. In summary, the on-line scheduling algorithm can produce schedules that achieve different objectives such as the minimal mean access time, the minimal variance of access time, the maximal service ratio and the minimal mean tuning time. We also evaluate a scheduling algorithm that can trade mean access time for low variance.

## Appendix: Deriving the Mean Access Time and the Variance of Access Time

In section 2, we define access time $t$ of a request as the duration time from when the request is made until the desired item appears on the broadcast channel. Let us define two random variables here: $T$, the issue time of a request, and $I$, the requested item in the request. $I$ is a discrete random variable taking integer values from 1 to M. The probability of item $i$ being requested is $p_i$. So

$$Prob[I = i] = p_i$$

If a request for item $i$ arrives at time $T$, its access time $t$ falls in the range $(0, s_i]$ depending on where $T$ resides between two consecutive broadcasts of item $i$. Since we assume that request arrival is governed by a Poisson process, the request arrives equally likely at any time. In the case of item $i$ being requested, $t$ is uniformly distributed over $(0, s_i]$ and the probability density function of $t$, $q_i(t)$, is:

$$q_i(t) = \begin{cases} \frac{1}{s_i} & , 0 < t \leq s_i \\ 0 & , \text{otherwise} \end{cases}$$

Since $t$ is a continuous random variable, cumulative distribution function for $t$ is obtained as:

$$F_i(x) = P[t \leq x | I = i] = \int_{-\infty}^{x} q_i(t)dt, \quad x \quad real$$

where $F_i(x)$ is the cumulative distribution function for $t$ given that $I = i$.

Using the *Multiplication Rule*, we may have the cumulative distribution function $F(x)$ for t:

$$\begin{aligned} F(x) &= P[t \leq x] \\ &= \sum_{i=1}^{M} (Prob[I = i] Prob[t \leq x | I = i]) \\ &= \sum_{i=1}^{M} (p_i F_i(x)) \end{aligned}$$

Let $g(t)$ be the probability density function of random variable $t$. Since $g(t) = \frac{dF(t)}{dt}$ and $q_i(t) = \frac{dF_i(t)}{dt}$, it follows that,

$$g(t) = \sum_{i=1}^{M} p_i q_i(t) \tag{9.A.1}$$

Now, we are able to derive the expressions for $\mu$ and $\sigma^2$. $\mu$ is the expected value of $t$. By the definition of expected value, we have

$$\begin{aligned}
\mu &= \int_0^\infty t g(t) dt \\
&= \int_0^\infty (t \sum_{i=1}^{M} (p_i q_i(t))) dt \\
&= \int_0^\infty \sum_{i=1}^{M} p_i (t q_i(t)) dt \\
&= \sum_{i=1}^{M} p_i \int_0^\infty t q_i(t) dt \\
&= \sum_{i=1}^{M} p_i \int_0^{s_i} \frac{t}{s_i} dt \\
&= \frac{1}{2} \sum_{i=1}^{M} s_i p_i
\end{aligned}$$

$\sigma^2$ is the expected value of random variable $(t-\mu)^2$. So, we have

$$\begin{aligned}
\sigma^2 &= \int_{-\infty}^{\infty} (t-\mu)^2 g(t) dt \\
&= \int_{-\infty}^{\infty} (t-\mu)^2 \sum_{i=1}^{M} (p_i q_i(t)) dt \\
&= \int_{-\infty}^{\infty} \sum_{i=1}^{M} (p_i (t-\mu)^2 q_i(t)) dt \\
&= \sum_{i=1}^{M} \int_{-\infty}^{\infty} p_i (t-\mu)^2 q_i(t) dt \\
&= \sum_{i=1}^{M} \int_0^{s_i} p_i (t-\mu)^2 \frac{1}{s_i} dt \\
&= \sum_{i=1}^{M} \frac{p_i}{s_i} [\frac{1}{3} (t-\mu)^3 |_0^{s_i}] \\
&= \frac{1}{3} \sum_{i=1}^{M} p_i s_i^2 - (\sum_{i=1}^{M} p_i s_i) \mu + (\sum_{i=1}^{M} p_i) \mu^2
\end{aligned}$$

Since $\mu = \frac{1}{2}\sum_{i=1}^{M} s_i p_i$ and $\sum_{i=1}^{M} p_i = 1$, the above equation can be simplified as

$$\sigma^2 = \frac{1}{3}\sum_{i=1}^{M} p_i s_i^2 - \mu^2$$

# Appendix: Minimizing the Variance of Access Time

$\sigma^2$ is a multi-variable function of $s_1, s_2, \cdots, s_M$. However, only $M - 1$ of the $s_i's$ can be changed independently instead of $M$. Let us define the share of bandwidth that each item occupies. Item $i$ is broadcast once every $s_i$ time period and each transmission takes $l_i$ time. So, the percentage of time taken by item $i$ during the broadcast is $\frac{l_i}{s_i}$. To utilize the bandwidth of broadcast channel to its full extent, we should make

$$\frac{l_1}{s_1} + \frac{l_2}{s_2} + \ldots + \frac{l_{M-1}}{s_{M-1}} + \frac{l_M}{s_M} = 1$$

or

$$s_M = l_M\left(1 - \frac{l_1}{s_1} - \frac{l_2}{s_2} - \ldots - \frac{l_{M-1}}{s_{M-1}}\right)^{-1} \tag{9.B.1}$$

Back to our objective of minimizing the $\sigma^2$, we have to find the schedule vector which makes $\frac{\partial\sigma^2}{\partial s_i} = 0, \forall i$. We now solve these equations, beginning with $0 = \frac{\partial\sigma^2}{\partial s_1}$.

$$0 = \frac{\partial\sigma^2}{\partial s_1} = \frac{\partial}{\partial s_1}\left[\frac{1}{3}\sum_{i=1}^{M} p_i s_i^2 - \left(\frac{1}{2}\sum_{i=1}^{M} p_i s_i\right)^2\right]$$

$$= p_1\left[\frac{2}{3}s_1 - \frac{1}{2}\left(\sum_{i=1}^{M} p_i s_i\right)\right] + p_M\left[\frac{2}{3}s_M - \frac{1}{2}\left(\sum_{i=1}^{M} p_i s_i\right)\right]\frac{\partial s_M}{\partial s_1} \tag{9.B.2}$$

From Equation 9.B.1, it can be found that

$$\frac{\partial s_M}{\partial s_1} = -\frac{s_M^2}{s_1^2}\cdot\frac{l_1}{l_M}$$

By substitution, Equation 9.B.2 becomes

$$0 = p_1\left[\frac{2}{3}s_1 - \frac{1}{2}\left(\sum_{i=1}^{M} p_i s_i\right)\right] - \frac{p_M s_M^2}{s_1^2}\cdot\frac{l_1}{l_M}\left[\frac{2}{3}s_M - \frac{1}{2}\left(\sum_{i=1}^{M} p_i s_i\right)\right]$$

which implies that

$$\frac{p_1 s_1^2}{l_1}\left(\frac{2}{3}s_1 - \frac{1}{2}\sum_{i=1}^{M} p_i s_i\right) = \frac{p_M s_M^2}{l_M}\left(\frac{2}{3}s_M - \frac{1}{2}\sum_{i=1}^{M} p_i s_i\right)$$

Similarly,

$$\frac{p_2 s_2^2}{l_2}\left(\frac{2}{3}s_2 - \frac{1}{2}\sum_{i=1}^{M} p_i s_i\right) = \frac{p_M s_M^2}{l_M}\left(\frac{2}{3}s_M - \frac{1}{2}\sum_{i=1}^{M} p_i s_i\right)$$

$$\cdots$$

$$\frac{p_{M-1}s_{M-1}^2}{l_{M-1}}(\frac{2}{3}s_{M-1} - \frac{1}{2}\sum_{i=1}^{M} p_i s_i) = \frac{p_M s_M^2}{l_M}(\frac{2}{3}s_M - \frac{1}{2}\sum_{i=1}^{M} p_i s_i)$$

In other words,

$$\begin{aligned} \frac{p_1 s_1^2}{l_1}(\frac{2}{3}s_1 - \frac{1}{2}\sum_{i=1}^{M} p_i s_i) &= \frac{p_2 s_2^2}{l_2}(\frac{2}{3}s_2 - \frac{1}{2}\sum_{i=1}^{M} p_i s_i) \\ &= \ldots \\ &= \frac{p_M s_M^2}{l_M}(\frac{2}{3}s_M - \frac{1}{2}\sum_{i=1}^{M} p_i s_i) \end{aligned}$$

This is equivalent to saying that

$$\frac{p_i s_i^2}{l_i}(\frac{2}{3}s_i - \frac{1}{2}\sum_{i=1}^{M} p_i s_i) = constant, \forall i, 1 \le i \le M$$

or

$$\frac{p_i s_i^2}{l_i}(\frac{2}{3}s_i - \mu) = constant, \forall i, 1 \le i \le M$$

## Appendix: Deriving the Service Ratio

A request for item $i$, arriving at time $t$, will be satisfied by the next transmission of item $i$, provided that the request does not leave before that transmission begins. Thus, the expected number of requests satisfied by a transmission of item $i$, denoted as $N_{s_i}$, can be obtained as follows:

$$\begin{aligned} N_{s_i} &= \int_0^{s_i} \lambda_i e^{-\tau(s_i - t)} dt \\ &= \frac{\lambda_i}{\tau}(1 - e^{-\tau s_i}) \\ &= \frac{p_i \lambda}{\tau}(1 - e^{-\tau s_i}) \end{aligned}$$

Note that the above expression calculates the number of those requests which arrive in the $s_i$ interval preceding a given transmission of item $i$, but do not leave until the given transmission begins. Due to the exponential model, the probability that a request will wait for interval $(s_i - t)$ or longer is $e^{-\tau(s_i - t)}$.

If the length of entire broadcast schedule is $C$, then the number of broadcasts of item $i$ is $n_i = C/s_i$, and the expected number of requests served by the schedule would be $\sum_{i=1}^{M} n_i N_{s_i}$. Also, the expected number of requests arriving in time period $C$ is $\lambda C$ for a Poisson process with rate $\lambda$. Then, we obtain the expected service ratio $R$ as

$$R = \frac{\sum_{i=1}^{M} n_i N_{s_i}}{\lambda C} = \sum_{i=1}^{M} \frac{p_i}{\tau s_i}(1 - e^{-\tau s_i}) \qquad (9.C.1)$$

# Appendix: Maximizing the service ratio

Expected service ratio $R$ is a multivariable function of $s_1,s_2,\ldots,s_M$. Since the fraction of bandwidth allocated to item $i$ is $l_i/s_i$, and the $M$ items together consume the available bandwidth, the values of the $M$ variables $s_1,s_2,\ldots,s_M$ are subject to the following constraint:

$$\frac{l_1}{s_1}+\frac{l_2}{s_2}+\ldots+\frac{l_{M-1}}{s_{M-1}}+\frac{l_M}{s_M}=1 \tag{9.D.1}$$

To solve the problem of finding maximum of $R$ subject to the above constraint, we use the method of Lagrange's multipliers. We consider function $R'$ below, where $\phi$ is an unknown constant.

$$\begin{aligned} R' &= R-\phi(\frac{l_1}{s_1}+\frac{l_2}{s_2}+\ldots+\frac{l_M}{s_M}-1) \\ &= \left[\sum_{i=1}^{M}\frac{p_i}{\tau s_i}(1-e^{-\tau s_i})\right] - \phi(\frac{l_1}{s_1}+\frac{l_2}{s_2}+\ldots+\frac{l_M}{s_M}-1) \end{aligned}$$

Next, we set all derivatives of $R'$ to 0. That is,

$$\frac{\partial R'}{\partial s_1}=\frac{p_1}{\tau s_1^2}(\tau s_1 e^{-\tau s_1}+e^{-\tau s_1}-1)-\frac{\phi l_1}{s_1^2}=0 \tag{9.D.2}$$

$$\frac{\partial R'}{\partial s_2}=\frac{p_2}{\tau s_2^2}(\tau s_2 e^{-\tau s_2}+e^{-\tau s_2}-1)-\frac{\phi l_2}{s_2^2}=0 \tag{9.D.3}$$

$$\ldots$$

$$\frac{\partial R'}{\partial s_M}=\frac{p_M}{\tau s_M^2}(\tau s_M e^{-\tau s_M}+e^{-\tau s_M}-1)-\frac{\phi l_M}{s_M^2}=0 \tag{9.D.4}$$

$$\frac{\partial R'}{\partial \phi}=\frac{l_1}{s_1}+\frac{l_2}{s_2}+\ldots+\frac{l_M}{s_M}-1=0 \tag{9.D.5}$$

Solving these equations to find the extremum point is difficult, but we can exploit their symmetry to obtain an useful result. From Equation (9.D.2), we get

$$\tau\phi=\frac{p_1}{l_1}(\tau s_1 e^{-\tau s_1}+e^{-\tau s_1}-1)$$

Similarly, we have

$$\tau\phi=\frac{p_2}{l_2}(\tau s_2 e^{-\tau s_2}+e^{-\tau s_2}-1)$$

$$\ldots$$

$$\tau\phi=\frac{p_M}{l_M}(\tau s_M e^{-\tau s_M}+e^{-\tau s_M}-1)$$

Denoting $\phi\tau$ as a constant $K$, it follows that

$$\frac{p_i}{l_i}(\tau s_i e^{-\tau s_i}+e^{-\tau s_i}-1)=K, \quad i=1,2,\ldots,M \tag{9.D.6}$$

The equation above gives a necessary condition that must be satisfied by the schedule vector at the extremum point where $R$ is maximized.

Before we show that this is also a sufficient condition for the optimal point, we first prove that there is only one solution of $s_1, s_2, \ldots, s_M$ satisfying the condition. Suppose that there are two solutions which are $s_1^a, s_2^a, \ldots, s_M^a$ and $s_1^b, s_2^b, \ldots, s_M^b$ and they satisfy the following equations, respectively.

$$\frac{p_i}{l_i}(\tau s_i^a e^{-\tau s_i^a} + e^{-\tau s_i^a} - 1) = K1, \quad i = 1, 2, \ldots, M$$

and

$$\frac{p_i}{l_i}(\tau s_i^b e^{-\tau s_i^b} + e^{-\tau s_i^b} - 1) = K2, \quad i = 1, 2, \ldots, M$$

Since $f(x) = \tau x e^{-\tau x} + e^{-\tau x} - 1$ is a monotonically decreasing function of $x$ for $x > 0$, if $K1 = K2$, then $s_i^a = s_i^b$, $\forall i$. Thus, if $K1 = K2$, the two solutions must be identical.

Now without loss of generosity, assume that $K1 > K2$. Thus, we get the following inequality.

$$\tau s_1^a e^{-\tau s_1^a} + e^{-\tau s_1^a} - 1 > \tau s_1^b e^{-\tau s_1^b} + e^{-\tau s_1^b} - 1$$

Therefore, $s_1^a < s_1^b$ must be true because the function $f(x) = \tau x e^{-\tau x} + e^{-\tau x} - 1$ is a monotonically decreasing function when $x > 0$. Similarly, we have $s_2^a < s_2^b, \ldots, s_M^a < s_M^b$. However, both solutions must satisfy constraint 9.D.1 and, obviously this cannot be the case if $s_i^a < s_i^b$, $\forall i$. In other words, there is only one valid solution for the set of equations obtained above, that achieves either a maximum or a minimum of $R$. By intuition, we know that a maximum value of $R$ exists, and is bounded by 1. Therefore, if a solution happens to be found satisfying the condition given in Equation 9.D.6, it must be the optimal point.

# Appendix: Deriving The Mean Tuning Time

Examine a request for item $i$ arriving between two consecutive transmissions of item $i$. Since the request adjourn time follows exponential distribution, the probability that the user waits for $w$ time units or longer is $e^{-\tau w}$ (cumulative distribution function) and the probability density function for $w$ is $\tau e^{-\tau w}$. Then, the expected adjourn time of a request that arrives $t$ time units after most recent transmission of item $i$ $(0 \le t < s_i)$ can be obtained as

$$\begin{aligned} W_t &= (s_i - t)e^{-\tau(s_i - t)} + \int_0^{s_i - t} w\tau e^{-\tau w} dw \\ &= \frac{1}{\tau}(1 - e^{-\tau(s_i - t)}) \end{aligned}$$

The total waiting time experienced by all users with requests for item $i$ arriving between two consecutive transmissions of item $i$ is

$$W_{s_i} = \int_0^{s_i} \lambda_i W_t dt$$

As in the analysis for the expected service ratio $R$, the expected waiting time $W$ of all requests (satisfied and unsatisfied both) for all items can be expressed as

$$W = \frac{\sum_{i=1}^{M} n_i W_{s_i}}{\lambda C}$$

where $n_i$ is the number of broadcasts of item $i$. It turns out that

$$W = \sum_{i=1}^{M} \frac{p_i}{\tau^2 s_i}(\tau s_i - 1 + e^{-\tau s_i}) \quad (9.E.1)$$

$$= \frac{1}{\tau}(1 - R) \quad (9.E.2)$$

## References

Acharya, S., Franklin, M., and Zdonik, S. (1995). Dissemination-based data delivery using broadcast disks. *IEEE Personal Communication*, pages 50–60.

Ammar, M. H. and Wong, J. W. (1985). The design of teletext broadcast cycles. *Performance Evaluation 5(4)*, pages 235–242.

Ammar, M. H. and Wong, J. W. (1987). On the optimality of cyclic transmission in teletext systems. *IEEE Transactions on Communications*, (Vol. COM-35 No. 1):68–73.

Banerjee, S. and Lee, V. O. K. (1994). Evaluating the distributed datacycle scheme for a high performance distributed system. *Journal of Computing and Information*, 1(1).

Gescei, J. (1983). *The Architecture of Videotex Systems*. Prentice Hall.

Herman, G., Gopal, G., Lee, K. C., and Weinrib, A. (1987). The datacycle architecture for very high throughput. In *Proc. of ACM SIGMOD*.

Imielinski, T., Viswanathan, S., and Badrinath, B. R. (1994). Energy efficient indexing on air. In *International conference on Management of Data*, pages 25–36.

Imielinski, T., Viswanathan, S., and Badrinath, B. R. (1996). Data on the air - organization and access. *IEEE Transactions of Data and Knowledge Engineering*.

Su, C.-J. and Tassiulas, L. (1997). Broadcast scheduling for information distribution. In *Proc. of INFOCOM'97*.

Zdonik, Z., Alonso, R., Franklin, M., and Acharya, S. (1994). Are disks in the air, 'just pie in the sky?'. In *IEEE Workshop on Mobile comp. System*.

Chapter 10

# INFORMATION DISSEMINATION APPLICATIONS

Eddie C. Shek
*Vizional Technologies, Inc*

Son K. Dao
*HRL Laboratories, LLC*

Darrel J. Van Buer
*HRL Laboratories, LLC*

**Abstract**: Satellite networks have unique advantage in applications like rapid deployment, situation awareness, and emergency response. The satellite network is often paired with terrestrial wireless networks to form a hybrid satellite-wireless infrastructure to support user mobility [Zhang and Dao, 1996; Dao and Perry, 1996]. However, the mismatches in characteristics of satellite and terrestrial wireless networks must be handled properly to allow effective utilization of available bandwidth, and timely delivery of highly relevant information. We have developed an Intelligent Information Dissemination Services (IIDS) model to address this. We believe it is the right model to support the dissemination and maintenance of extended situation awareness throughout such a network information infrastructure in a seamless manner.

**Key words**: Hybrid satellite-wireless networks, information dissemination, user profile

# 1. INTRODUCTION

Situation awareness and emergency response are two important information-centered applications that require support for a large number of geographically distributed mobile users collaborating on a common mission and with interests in common situation domain. For example, after a natural disaster, emergency response teams and residents of affected areas need to share situation information (e.g., road condition) as well as their up-to-date locations in order to coordinate evacuation; at the same time, they also receive a large variety of multimedia information from outside source (e.g., real-time video from CNN, satellite sensor data from NASA, weather maps and forecasts) to enhance decision making. Shared situation awareness, during real-time mission execution, will be achieved by a hierarchical propagation of information throughout the operational organization. To effectively adapt and react to rapidly evolving scenarios, units at all levels of command must perceive an extended awareness of the situation and often act autonomously while remaining globally consistent in the overall mission objective. The rapid deployment, user mobility, and wide-area requirements strongly suggest dictate the use of a hybrid satellite-wireless networking infrastructure [Dao and Perry, 1996; Zhang and Dao, 1996] for such applications.

Satellite networking services based on geostationery-earth-orbit satellites (GEO) are widely available, while a number of services based on low-earth-orbit (LEO) and medium-earth-orbit (MEO) satellites are currently being planned. Because of GEO satellites' large footprints, they are particularly suitable for broadcast delivery of information to a large number of geographically distributed users. However, the high altitude of GEO satellites gives rise to high transmission latency (approximately 250ms one way). In addition, to achieve high bandwidth (tens to hundreds of Mbps), GEO satellite networks have to employ large uplink and downlink antenna that limits the mobility of satellite terminals. On the other end of the spectrum from satellite-based networks, terrestrial wireless networks (such as those based on 2G/3G cellular technologies) are highly mobile and have low transmission latency. However, their ranges and bandwidth are generally limited; a terrestrial wireless link can have bandwidth as low as tens of kbps and the range is as low as a few kilometers.

While neither satellite nor terrestrial wireless networking satisfies situation awareness and emergency response applications' need for wide-area support for mobile users, when combined appropriately they present a complete networking solution. In particular, the satellite link can be established for rapidly deployable high-bandwidth wide-area networking while local-area data redistribution from local satellite-terminal gateway is handled by a mobile wireless local area network. Nevertheless, such a network suffers from high end-to-end network latency, limited end-to-end bandwidth, as well as mismatched subnet characteristics.

Global mobile wireless network connectivity [Katz et al., 1996] is becoming readily available. Network-layer techniques such as TCP snooping [Balakrishnan el at., 1995] have been developed to improve networking efficiency over terrestrial wireless and satellite networks. However, we still need to develop innovative application-layer approaches to optimizing information services that run on top of such mobile wireless network infrastructures. In particular, we need to overcome the limitations of hybrid satellite-wireless networking infrastructures and to support the dissemination and maintenance of extended situation awareness throughout such a network information infrastructure in a seamless manner. We thus developed the Intelligent Information Dissemination Service (IIDS) model.

IIDS introduces technologies for proactive and adaptive reliable multicast-based information dissemination control from a hierarchy of dissemination proxies to users in the field at the "appropriate" rate, level of detail, format, and presentation. The main services provided by IIDS include:

- Dynamic User Profile Aggregation. Users' information needs are captured in profiles that specify data requirements and predicted change in future requirements. IIDS dynamically aggregates and clusters semantically similar profiles into hierarchy of user groups serviced by common multicast channels. Each packet in information stream is matched against the hierarchy to determine the list of recipient user groups.
- Predictive Push and Caching. To overcome the high end-to-end latency of a satellite-based network, IIDS anticipates users' incremental needs and proactively and seamlessly fill each client's cache with neighborhood situation information so that future requests can be satisfied by locally available data.
- Bandwidth-Aware Filtering. Because of the variation in bandwidth availability in subnets in a hybrid satellite-wireless network, application-level dataflow "throttling" is needed to filter and transform information packets to level of detail appropriate for disseminating to its user groups to alleviate congestion caused by dataflow from high-bandwidth to low-bandwidth subnet.
- Reliable Multicast-Based Data Dissemination. The scalable group communication model of IP multicast forms a natural basis for large-scale data dissemination. However, IP multicast is unreliable and an appropriate IP multicast based reliable group data distribution protocol

has to be used to effectively support dissemination of important information that demands absolute delivery reliability.

## 2. IIDS ARCHITECTURE

The Intelligent Information Dissemination Service employs a three-tier architecture (see *Figure 10-1*) composed of: (i) a source IIDS server at the satellite uplink center, (ii) a collection of intermediate IIDS proxies at the junction between the satellite network and local wireless networks, and (iii) IIDS clients running on user machines, each of which is served by a specific intermediate IIDS proxy. It realizes the proactive and adaptive information dissemination from an information source to a large number of mobile users in the field.

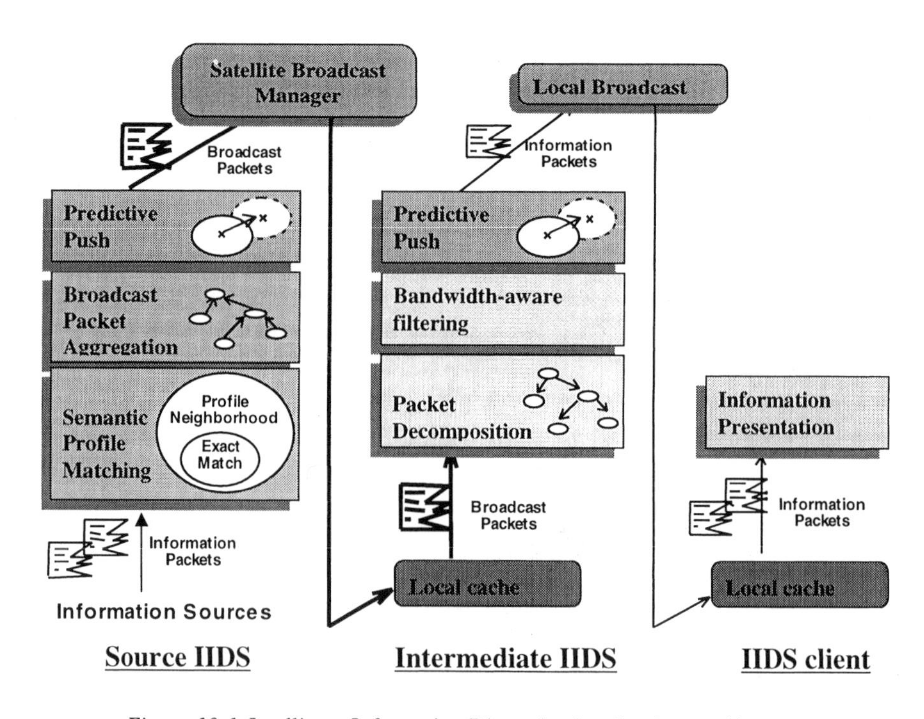


*Figure 10-1.* Intelligent Information Dissemination Services architecture

Broadcast data dissemination is most effective when each broadcast information packet has multiple interested, or receiving, parties. To maximize the value of multicast dissemination, IIDS proxy implements methods to collect similar user profiles into aggregate user classifications that are served by independent multicast channels of custom information packets. The IIDS server treats IIDS proxies in much the same way as how each IIDS proxy treats its users; it clusters the information profiles of IIDS proxies for wide-area multicast dissemination. Packets in information stream to each IIDS proxy are semantically matched against user hierarchy maintained by the proxy to determine how they are redistributed to local user groups.

Each client is associated with a specific intermediate IIDS proxy to which it registers the user's interest profile. It contains a local buffer that caches the part of the multicast dissemination stream that falls in the neighborhood of the user's interest domain. The intermediate IIDS proxy clusters its clients based on their interest profiles and matches dynamic situation information from the source IIDS server against its user profile hierarchy, and pushes matched information packets to the appropriate user groups over multicast channels. The profile of each user group created as a result of clustering is instantiated at the IIDS server or proxies as a "filter" that controls the flow of information to the appropriate multicast channels for the group. When the profile of a user changes, the update will be propagated through the proxy to the server, potentially changing the profile clustering at the proxy and the server as a result. Because of the relatively limited bandwidth of the wireless link between the proxy and the client (compared to that of the satellite link between the proxy and the server), the proxy may have to perform data transformation to reduce data size before re-distributing information streams to its clients. The source IIDS server connects to real-time multimedia information sources such as video-audio-text news feed and road map servers and packages related data into information packages disseminated to its client intermediate IIDS proxies. The relationship between the client and the proxy is very similar to that between the proxy and the server, with the intermediate IIDS proxy representing the aggregate interest of all its users.

## 3. MOBILE USER PROFILING

A mobile user operating in the field changes location, consume resources, investigate situations "on the horizon", and perform other *incrementally evolving activities*. A user's information needs are therefore continually evolving in a neighborhood of interrelated data centered on the user's profiles of primary information interests. Traditional profile-based information dissemination services (e.g., Internet-based music recommendation services, push-based data services such as PointCast) employ a static user profile that capture user interest as either a list of features that the user is interested in or as a class member of some classification. These profiles are then matched against statically created information dissemination channel.

The location of a vehicle-based mobile user is an integral part of the specification of the user's interest profile. In addition, under very many circumstances users move around in the real world following well defined natural and/or artificial routes such as roads, rivers, tracks, air channels, etc. These routes limit the areas to which a user is accessible and hence the geographic coverage of a user's interest is often limited to areas along these routes. A simple and common way of modeling an interest in information pertaining to a geographic region is by means of a rectangular bounding box. While convenient, information requests modeled as rectangular bounding boxes can be inaccurate, especially for that of a group of clustered users each of whose interest is limited by accessibility to area reachable by road. Instead, we divide the entire geographic area of interest is divided into a rectangular grid and the geographic area of interest of each mobile user $u$, denoted as *Profile(u)*, is defined as the set of grids that lie in its neighborhood along reachable paths that the user may arrive at in the near future. *Figure 10-2* shows 3 objects ($a$, $b$, and $c$ each represented by a circle) and their respective neighborhoods (represented by the set of small squares) are depicted. The profile of a group of clustered user group $G$, denoted as *Profile(G),* is similarly the union of all grids in the group members' profiles. More formally, for user group $G$ containing users $u_1, \ldots, u_k$,

$$Profile(G) = Profile(u_1) \cup \ldots \cup Profile(u_k).$$

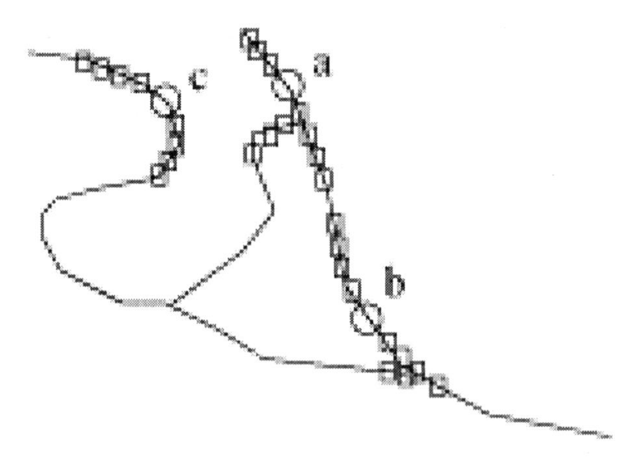


*Figure 10-2.* Neighborhoods of moving entities representing the geographic coverage of their interests

Mobile users' locations and paths are mapped onto a cartographic representation of geographic information that allows reasoning. The cartography needs to be modeled by a data structure that can guarantee an efficient computation of the distance between each pair of objects since the clustering objective function is based upon such a distance. Spatial networks [Shekhar et al., 1996] (alternatively called Spatial Graph) have been exploited in many different domains as the kernel data representation; examples of these domains are: transportation systems, air traffic management, urban management, as well as all the different types of utility networking such as power, telephone, water, and gas. A spatial network is a richly connected graph where nodes are labeled with a geographical *(x,y)* location of the specific entity to be modeled, and edges link pairs of entities. As an example, a spatial graph for a road transportation system would be modeled by having a node for each road intersection and an edge for each road segment connecting two intersections. Nodes would be labeled with Euclidean *(x,y)* coordinate of the intersection, while edges would be associated to various information depending on the target application (e.g., length of the road segment, name of the street, type of street, etc.).

In order to closely capture the dynamic nature of mobile users, we modified the spatial network model to support a finer grained route representation. Specifically, our spatial graph representing a route network contains (i) a node for each route vertex, (ii) a node for each road intersection, and (iii) most importantly, intermediate nodes that are created to break long straight road into segments shorter than a certain threshold (i.e., *MAX_EDGE_LENGTH*) defined by the user. A spatial graph in our system is an approximation of the coordinate space in the input map for all the points belonging to a route in the map. The approximation ratio is controlled by the parameter *MAX_EDGE_LENGTH* that is set by the user. The value of *MAX_EDGE_LENGTH* depends upon the current application and the accuracy of the related equipment (e.g., GPS tolerance).

Once the spatial graph is created it can be used at run-time to dynamically compute distance among mobile users. The coordinates of each moving entity are mapped into the closest node in the spatial graph. A neighborhood of a mobile user is the set of nodes in the spatial graph that are within a threshold distance from the user's location. The more grids overlap between the two neighborhoods, the (spatially) closer we consider the two users are. This measure of vicinity is chosen rather than the Euclidean distance between the locations of users since the actual distance between pairs of entities moving along defined paths rarely coincides with their Euclidean distance. The adopted graph based representation discussed above allows an easy and efficient computation of the vicinity of user profiles by means of a conventional linear-time graph visit algorithm on the spatial graph [Weiss, 1996].

# 4. DYNAMIC USER PROFILE CLUSTERING AND AGGREGATION

Each class in a user clustering represents a group request that is an aggregation of a collection of similar user requests. A user clustering is a hierarchy of group requests representing increasingly specialized range as one goes down the hierarchy. IIDS server performs broadcast packet aggregation operation that collects related information updates into broadcast packets based on clustering of user profiles (e.g., group road condition maps covering overlapping areas into a single larger map before multicasting to users requesting the maps), It aims to maximize number of users interested in each broadcast packet and reduces data size for more effective bandwidth utilization. The submission or update of a user profile results in the assignment of the request to a class in the profile hierarchy; at the same time, classes may be created, merged, split, and deleted, while users' requests may get reassigned to other classes during user clustering discovery. A user is notified of changes when its class assignment, and hence data delivery multicast channel, changes.

User profile clustering belongs to a class of conceptual clustering problems that have been studied extensively by the artificial intelligence community as a means to extract hidden inference information [Fisher, 1987], and by the information retrieval community as a method to improve the quality of query results [Charikar et al., 1997]. Clustering algorithms (e.g., COBWEB [Fisher, 1987], CLUSTER/2 [Michalski and Stepp, 1983], and AUTOCLASS [Cheeseman et al., 1988]) generally share the common goal of discovering structure in data but differ in the objective function used to evaluate clustering quality and the control strategy used to search the space of clustering.

Our algorithm to generate profile clustering to allow effective information dissemination differs from existing conceptual algorithms in that the goal of the clustering is more than only to optimize the accuracy of clustering. In addition to accuracy in the clustering, also important in incremental profile clustering are simplicity (small number of groups and shallow graph), and stability (infrequent reassignment of user requests to groups). We desire a simple clustering since networks generally have a practical limit on the number of multicast channels that can be supported and that the overhead of information packet matching is dependent on the number of groups in the clustering. At the same time, it is important for the profile clustering to not change frequently because a large number of users are potentially interrupted when the clustering changes, while re-materializing a changed clustering in the form of new multicast channels can be an expensive process. Our algorithm alleviates the problem of instability by sacrificing some of the optimality in the clustering through the use of a request group cover diameter tolerance. In addition, we allow the bounding of the number of groups that translates to the number of multicast channels that is limited in a real network. The algorithm is composed of 2 components: (i) incremental clustering that adds new user requests into existing clustering, and (ii) a continuous process of adaptive re-clustering that reconfigures the existing clustering which optimality may have deteriorated since its creation as users move and change interest over time.

The dynamic clustering framework can work with arbitrary application-dependent user profile formulations. To simplify discussion on clustering, we assume that each user's information needs is determined entirely by his geographic location and we capture the profile of each mobile user as a collection of grids lying in its neighborhood along reachable paths as described earlier.

## 4.1 Incremental Clustering Framework

The submission, update, or cancellation of user requests triggers the profile clustering process. Given the needs of profile clustering for intelligent vehicle information sharing, the algorithm to reconfigure an existing profile clustering when a new request is submitted consists of 3 main steps:

1. Try to find an existing group in the clustering that closely covers the request. A profile group $G$ covers a user request $u$ if the group request assumed by the group completely subsumes the request. It is often desirable to include the new user profile $u$ into the group $G$ since the information carried in the multicast channel servicing $G$ is guaranteed to satisfy the information needs of $u$. However, if $G$ completely covers $u$ but the collection of grids in the profile of $G$ is much larger than that of $u$, we may not want to include u into group $G$ since doing so means user $u$ will be receiving information satisfying the profile $G$ which contains a lot of material irrelevant to $u$. As a result, a coverage threshold, *COVER_THRESHOLD*, is defined as a means to control the inaccuracy of clustering that can be tolerated. A request is assigned to a group that completely covers it only if the difference between the their coverage is below the threshold. For example, if the threshold is small, a request is assigned to an existing group only if their coverage is very similar. On the other hand, if the threshold is large, a request may be assigned to an existing group in the profile clustering even if there is little overlap. If multiple existing groups satisfy the criteria, we cluster the user into the group that has the smallest coverage to get the tightest fit for the user.

More formally, given our spatial formulation of user profiles, we cluster user $u$ into group $G$ if and only if

a) $Profile(G) \supseteq Profile(u)$,

b) $|Profile(G) - Profile(u)| \,/\, |Profile(G)| < COVER\_THRESHOLD$, and

c) there does not exist group F such that $|Profile(F)| < |Profile(G)|$.

2. Try to expand existing group that is the closest in coverage to the request. If we cannot simply include a user into an existing group that completely cover it, we may want to consider clustering the user $u$ into an existing group $G$ which coverage largely overlap that of the user and expanding the group's coverage in the process. However, we may not want to do so if doing so would cause the coverage of the resultant group to expand significantly to the extent that members of the group begin to receive much irrelevant information. As a result, an expansion threshold, *EXPAND_THRESHOLD*, is defined to control the extent of expansion in a group's coverage that is allowed. Specifically, the group that is closest to the request in coverage cannot be expanded to include the request if the increase in the group's coverage from its original coverage is larger than the threshold. If the threshold is small, groups are allowed very limited expansion in coverage. On the other hand, groups do not expand arbitrarily if the threshold is large. The threshold introduces a tunable means of allowing inaccuracy in clustering as a tradeoff for profile clustering stability. If multiple existing groups satisfy the criteria, we cluster the user into the group that has the smallest expanded coverage to minimize the deterioration of clustering quality for existing users. More formally, given our spatial formulation of user profiles, we cluster user $u$ into group $G$ and expand the coverage of $G$ if and only if

a) $|Profile(G) \cup Profile(u) - Profile(G)| \,/\, |Profile(G)| < EXPAND\_THRESHOLD$, and

b) there does not exist group F such that

$$|Profile(F) \cup Profile(u)| < |Profile(G) \cup Profile(u)|.$$

3. If all else fails, generate a new profile group that covers the user request. The new group will serve as a new locus of clustering to which new user requests are attracted. Moreover, it may later be merged with other groups as its coverage migrates due to changes in its membership and the coverage of its members.

The optimal choices of the configurable parameters *COVER_THRESHOLD* and *EXPAND_THRESHOLD* are dependent on the application requirements and the characteristics of network infrastructure. This dependency will be illustrated later when we present experiment simulation results. Note also that the evaluation of coverage depends on the formulation of user profiles. The formulation of the decision rules may be different if we choose a different profile formulation, but the general framework remains the same.

## 4.2 Adaptive Re-clustering

Once the objects are clustered, we try to preserve the original clusters as far as possible - as objects move or require new information, the group they are in is modified to reflect these changes. However, over time this may lead to improper clustering of entities. The entities in a particular group may drift far apart enough from each other such that the coverage area of the group becomes undesirably large. Also, two clusters may change their coverage such that their information coverage overlaps to an extent that it would be justified to merge the two clusters into a single cluster. The problem of clustering evolving user profile poses interesting problems. If the re-cluster frequency is high, it becomes expensive and intrusive since it may result in frequent changes in assignment of user profiles to groups. On the other hand, infrequent re-clustering allows the quality of clustering to deteriorate as user profiles change. This may result in a higher network load, and imposes a higher caching resource requirement. As a result, the main issue is to strike a balance between avoiding frequent re-clustering and achieving accurate clustering. Factors affecting the optimal time frame to consider include the available bandwidth, cache size, the expected update rate of information, and the expected rate of change of user profiles. Later in this section we will show how the choice of re-clustering frequency affects the clustering quality.

We tackle the dynamics of user profiles by using a two-phased approach towards maintaining group assignments. Periodically, the assignment of entities to clusters is "reviewed". This review consists of two phases: a splitting phase and a merging phase:

- *Splitting Phase*. In the splitting phase, each group and the entities in it are examined and if the entities in the group have moved significantly away from each other, the group is split into multiple smaller clusters. We tackle the task of deciding whether a group should be split into multiple clusters by re-clustering the entities in the group within themselves. If the entities in the group should logically belong to a single cluster, the re-clustering will result in a single cluster. If, on the other hand the entities in the group have moved far apart from each other since the last time the clusters were reviewed, so much so that their being in the same group cannot be justified, the result of the re-clustering will be two or more smaller clusters consisting of subsets of the entities in the original cluster. Re-clustering the entities within a group is less expensive because the number of entities being re-clustered is a lot smaller than the total number of entities in the system. Within a group, we first randomly pick an entity as the locus of clustering. All other entities in the group are then incrementally clustered in a random order following the incremental clustering algorithm described before.

- *Merging Phase*. In the merging phase, groups in the existing clustering are pair-wise examined. To simplify the clustering and reduce the number distinct groups, we may want to merge 2 groups if they overlap to a large extent. Merging 2 groups that do not sufficiently overlap results in a group that contains members whose coverage does not closely fit the group coverage and this results in the member users subscribing to a multicast data channel in IIDS that contains a lot of irrelevant information. As a result, for each pair of groups in a clustering, we measure of the ratio of grids that appear in only one group against the total number grids. If this ratio is less than the parameter *GROUP_MERGE_THRESHOLD*, the clusters are merged if no other pairs of groups have a lower ratio. As the threshold increases, the less overlap 2 groups have to have for them to be merged and the more relaxed the merging criteria is. The merging process repeats until no pairs of groups satisfy the merging criteria. More formally, given our formulation of user profiles, we define the "group merge expansion ratio" as gmer(F,G) = $|(Profile(F) \cup Profile(G)) - (Profile(F) \cap Profile(G))| / |Profile(F) \cup Profile(G)|$, and we merge 2 groups *F* and *G* if and only if (i) $gmer(F,G) < GROUP\_MERGE\_THRESHOLD$ and (ii) there does not exists other groups *I* and *J* such that $gmer(I,J) < gmer(F,G)$.

## 4.3 Evaluation

The profile-clustering algorithm contains a number of parameters that can be adjusted to tune the balance between the overhead, accuracy, and simplicity of the clustering generated. They include the re-clustering frequency, the cluster-merging threshold, and the group-joining threshold. One of the most important parameters that can be used to adjust the quality of clustering is the group-merging threshold that controls the allowable overlap between overlapping groups before they are combined into an aggregated group. Here, we will show some simulation results and discuss the effects of changing one of the parameters, namely the cluster merging threshold, on vehicle information dissemination. We will present a series of 15-minute simulation runs of the dynamic clustering algorithm with a simulated user environment consists of 30 vehicles following random paths in the road map shown in *Figure 10-2*.

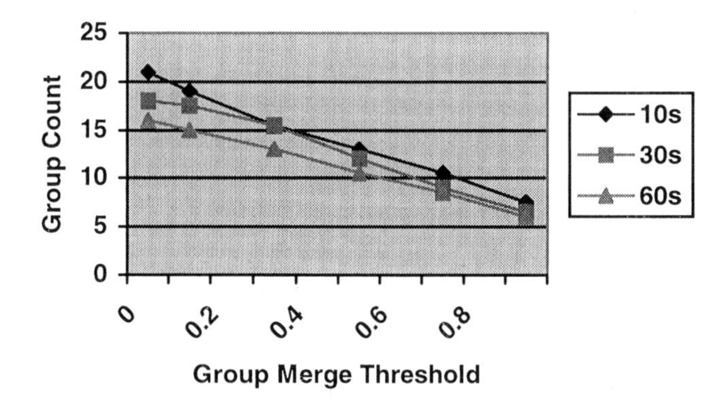


*Figure 10-3.* Plot of group count against group merging threshold and re-clustering period

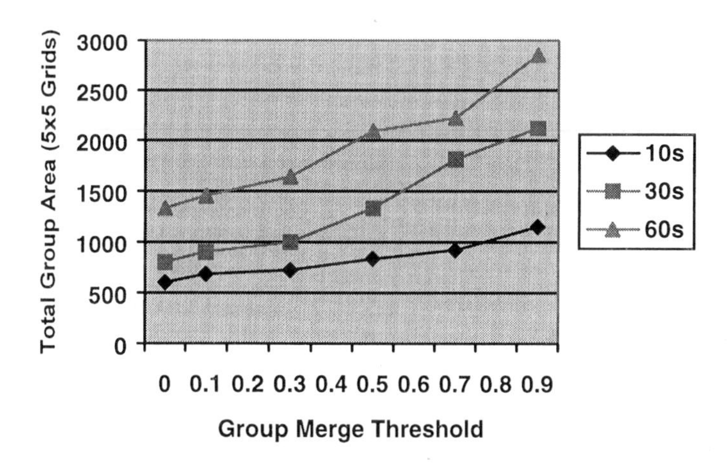


*Figure 10-4.* Plot of total group area against group merging threshold and re-clustering period

*Figure 10-3* and *Figure 10-4* show the effects of re-clustering parameters on the clustering quality. Specifically, *Figure 10-3* shows a plot of the average number of groups in the clustering for different group merging thresholds and re-clustering periods (with *COVER_THRESHOLD* of 0.1 and *EXPAND_THRESHOLD* of 0.1). For any given re-clustering frequency, the number of groups in the clustering decreases as the group-merging threshold increases since groups are allowed to be merged only with a decreasing amount of overlap during the re-clustering process. At the same time, for a given *GROUP_MERGE_THRESHOLD*, the group count in the clustering decreases slowly as the re-clustering period increases. This can be mainly attributed to the fact that existing groups in a clustering are allowed to increased expansion in coverage as the re-clustering period increases since users in groups move. The expanded coverage of groups resulted in increased likelihood that they overlap significantly and hence merged during the re-clustering process. Nonetheless, this effect is partially countered by the splitting of large groups during the splitting phase of re-clustering.

*Figure 10-4* shows a plot of the total area covered by the groups for different group merging thresholds and re-clustering periods. In this case, each group includes the coverage of its members and all 5x5 unit grids along the map-graph connecting them. Assuming a constant geographic information density, i.e., that approximately the same amount of data are being generated by mobile users and other data collection centers to describe the situation at each area, a plot showing the network bandwidth requirement will have the same shape as *Figure 10-5* for varying group merging thresholds. This information can be used, with the multicast channel limit, to determine the appropriate group-merging threshold to use given different network bandwidth availability and allocation. The simulation results show that for any given re-clustering period, the total coverage size increases as the group merging threshold increases and group count decreases, since increase aggregation causes more grids "between" users to be included in groups. At the same time, for any given *GROUP_MERGE_THRESHOLD*, the total coverage size increases significantly as the re-clustering period increases despite a relative small increase in group count. This implies that the average size of each group's coverage increases significantly which points to ineffective bandwidth utilization as IIDS data dissemination channels can be expected to contain an increasing portion of information that is irrelevant to any individual user. While it seems that it is best to re-cluster frequently, it is important to note that re-clustering is an expensive process and frequent re-clustering may impose a heavy computation load on IIDS servers and proxies when the number of users is large.

Note that since each group maps into a multicast channel in our information dissemination system, a limit on the number of allowed multicast channels in the networking infrastructure imposes a limit on the maximum group-merging threshold. For example, if the network infrastructure limits the number of multicast channels to 12, *Figure 10-3* shows that the group merging threshold should be set to at least 0.55 if the re-clustering threshold is 10 seconds given that all the other parameters remain the same. A group-merging threshold of 0.55 as dictated by a multicast channel limit of 12 results in a total group coverage of 870 units as shown in *Figure 10-4*.

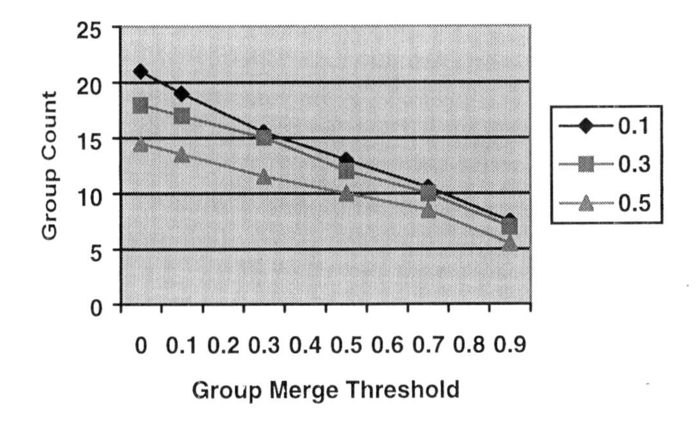


*Figure 10-5.* Plot of group count against group merging threshold and expansion threshold

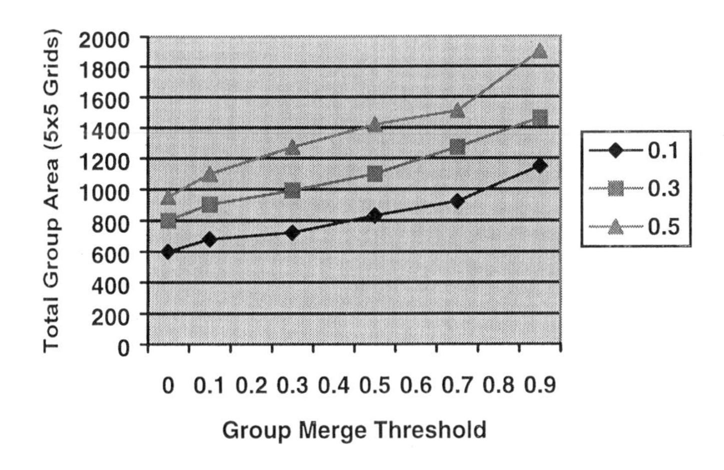


*Figure 10-6.* Plot of total group area against group merging threshold and expansion threshold

*Figure 10-5* and *Figure 10-6* show the effects of the parameters controlling group merging, namely *GROUP_MERGE_THRESHOLD* that tunes the merging phase of the re-clustering process and *EXPAND_THRESHOLD* that controls the incremental clustering process as well as the splitting phase during re-clustering, have on clustering quality. *Figure 10-5* shows a plot of the average number of groups in the clustering for different group merging thresholds and *EXPAND_THRESHOLD*s (with *COVER_THRESHOLD* of 0.1 and re-clustering period of 10 seconds). Not unexpectedly, given an expansion threshold, we observe that the group count decreases as the group-merging threshold increases since groups are allowed to merge only with a decreasing amount of overlap. At the same time, given a *GROUP_MERGE_THRESHOLD*, the group count in the clustering decreases as the expansion threshold increases since it increases the likelihood that a new user profile will be merged into an existing group and it becomes less likely that an existing cluster will be split up during re-clustering.

*Figure 10-6* shows a plot of the total area covered by the groups for different group merging thresholds and expansion thresholds. Given a group merge threshold, the total coverage size increases as the group merging threshold increases and group count decreases, since more clustering causes more space "between" users to be included in groups both during incremental clustering and the splitting phase of re-clustering. And as expected, given an expansion threshold, the total coverage size as well as the average group coverage increases with the expansion threshold while the average group coverage size. This observation seems to advocate the choice of a small expansion threshold, but again the optimal choice may be constrained by the practical maximal limit on group count (and hence minimal limit on expansion threshold given a group merge threshold).

# 5. DATA DISSEMINATION TECHIQUES

## 5.1 Predictive Dissemination and Caching

Relevant information may not be delivered to a user group at a certain time because of the limited bandwidth of terrestrial wireless network. As a result, it is important that information in the neighborhood of a user's immediate interest coverage be cached in local buffer when bandwidth is available for their delivery so that they can be served to user in the future (even though they may be slightly up-of-date) from the local cache even if bandwidth is no longer available. This proactive dissemination of neighborhood information is achieved in IIDS through two mechanisms.

- First, the user profile may include information that it may need in the near future. For our formulation of user profile, this can be manifested as grids that the user may arrive at in the future. For non-time-sensitive information such as a road map, this makes it possible for IIDS to disseminate information to a user before the information is actually used.
- Second, it is a side-effect of clustering users with similar interest needs (e.g., in a geographic neighborhood) into a user group that listens to a

common multicast channel for data dissemination; any user will receive information that are destined for its fellow user group members even though the data may not perfectly matches the user's needs.

Dynamic user profile clustering forms a basis for providing support for proactive information dissemination that anticipates users information needs and seamlessly fill each client's cache with neighborhood, or likely to be relevant, situation information.

## 5.2 Bandwidth-Aware Filtering

Because of the difference in bandwidth availability in subnets (high-bandwidth satellite link vs. relatively low-bandwidth terrestrial wireless link), application-level dataflow "throttling" is needed to filter information packets to alleviate congestion caused by flow of data from high-bandwidth to low-bandwidth subnet. The clustering of users with similar profiles into a group that is serviced by a common multicast data dissemination channel goes a long way towards achieving the goal. However, it does not always completely solve the problem and as a result IIDS proxies implement a complementary bandwidth-aware filtering framework that allows more traditional filtering techniques to be supported to filter an information stream [Brooks, 1995; Fox et al., 1996; Fox and Brewer, 1996] into an appropriate level of detail, reduce the utilization of available bandwidth in data distribution channels, and reduce latency.

The basic premise for bandwidth-aware filtering at the intermediate IIDS proxy is the tradeoff between data quality and data delivery cost (and data size). For example, it is possible for the IIDS proxy to transform information into various representations of different sizes (e.g., audio and its text transcript, or an image in various resolutions and sizes) before redistribution control the reduction in data size trading data size for "compressing" data as bandwidth utilization increases. For image data, an intermediate IIDS proxy may lower the resolution, decrease the color depth, reduce the image size, or increase the compression ratio (and hence lossiness for lossy image representation). However, it is important to not ignore inherent application and data characteristics to maximize the utility of information after transformation. For example, a road map loose all useful detail after its resolution has been reduced significantly; in the case where bandwidth availability dictates that an image be transformed to a reduced-resolution version that falls under the acceptable resolution before dissemination, we are better off not sending the transformed image at all and save the bandwidth for the dissemination of some other piece of information over the same channel.

## 5.3 Reliable Multicast-based Dissemination

Users on a situation awareness application often are collaborating on the same mission and hence have closely related information needs. The commonality of information interest makes the IP multicast group communication model a natural basis for data dissemination in IIDS.

Many reliable multicast protocols have been developed to introduce reliability in data delivery to the unreliable UDP-based IP multicast [Floyd et al., 1995; Macker et al., 1996]. They differ in their implementations and hence their appropriateness over the hybrid satellite-wireless network infrastructure also differs. One dimension of classifying reliable multicast protocols is based on whether the client or the server has to be responsible for guaranteeing reliable data delivery as well as which is responsible for repair when data loss has been detected. Sender-reliable refers to schemes in which the responsibility for guaranteeing reliable data delivery falls on the sender which monitors positive acknowledgements from receivers and issues repair when error is detected; the scheme is simple and provides centralized control but is limited in scalability because of the possibility of ACK implosion and the sender's need to keep track of receiver states. On the other hand, receivers in receiver-reliable schemes are responsible for identifying data loss and signaling it through negative acknowledgements. Receiver-reliable schemes can be sub-classified based on how errors are repaired. The sender in sender-oriented scheme is responsible for accepting NACKS and re-send missing data, while receivers in receiver-oriented schemes caches data and repairs errors at receivers in the same multicast group. The scalability of receiver-reliable schemes may be limited in low quality networks due to NACK implosion. In addition, these receiver-oriented schemes impose a buffering requirement at receivers that may not be reasonable for receivers that own limited available resource.

In IIDS, we adopted the Multicast Dissemination Protocol (MDP) for reliable multicast-based information dissemination [Macker and Adamson, 1999] of data that requires a guarantee in delivery. IP multicast is still used for streaming audio and video for which reliability is not necessary and actually inappropriate because of timing constraints. MDP is a receiver-reliable sender-oriented scheme that implements a NACK suppression algorithm in which receivers listen to the multicast channel for a NACK multicasted by another receiver requested re-multicast of a data item if it recognizes the loss of the piece data. The algorithm allows NACKs to be amortized since multiple receivers often miss the same data items unless the data loss happens at the "last mile" to a particular receiver. MDP satisfies the multicast reliability requirements of a number of applications including situation awareness and image dissemination that have a large number of resource-limited users.

# 6. IMPLEMENTATION AND DEMONSTRATION

We have implemented the IIDS server and proxy supporting mobile user profiling, dynamic clustering, bandwidth-aware filtering, and MDP-based reliable multicast data dissemination as described earlier [Shek et al., 2000]. We have also demonstrated the software in a hybrid satellite-wireless network testbed called DWBN (Digital Wireless Battlefield Network) [Dao et al. 1999]. *Figure 10-7* shows the DWBN architecture: a seamless integration of three portions -- the backbone network (the Internet), a satellite network (11.8Mbps), and a cellular-based terrestrial wireless network (32kbps). *Figure 10-7* also shows how IIDS worked on this testbed network.

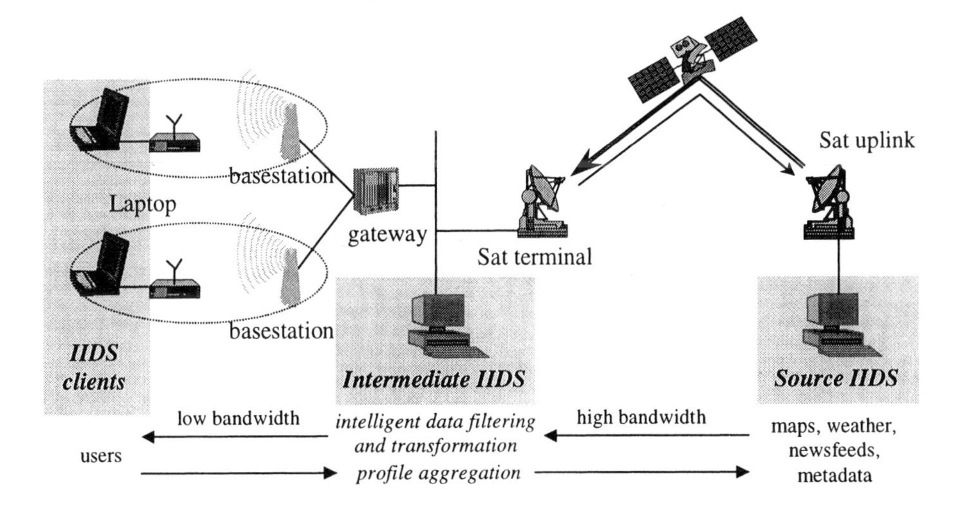


*Figure 10-7.* IIDS deployment in the Digital Wireless Battlefield Network (DWBN)

# 7. CONCLUSIONS

In this chapter, we described the design of Intelligent Information Dissemination Services (IIDS) that runs on a hybrid satellite-wireless mobile network and handles the hybrid network performance mismatches. It is built upon *user profile aggregation* that incrementally groups users into communities sharing common interests to enable multicast-based information dissemination. Based on the user grouping, *semantic profile matching* customizes information streams based on matching user group interest profiles. By taking into account of expected changes in user profiles, profile-oriented data dissemination achieves *predictive push and caching* that anticipates future user needs and minimizes latency of data request by making data available before they are explicitly requested. Finally, *bandwidth-aware filtering* adapts information streams to resource bandwidth availability to gracefully hide the bandwidth mismatch between the satellite and wireless links in the hybrid network infrastructure.

## REFERENCES

Balakrishnan, H., Seshan, S., and Katz, R. (1995). Improving Reliable Transport and Handoff Performance in Cellular Wireless Networks, *ACM Wireless Networks Journal, 1(4).*

Brooks, C., Mazer, M. Meeks, S., and Miller, J. (1995). Application-Specific Proxy Servers as HTTP Stream Transducers, In *Proceedings of the Fourth International World Wide Web Conference.*

Charikar, M., Chekuri, C., Feder, T., and Motwani, R. (1997). Incremental Clustering and Dynamic Information Retrieval, In *Proceedings of the Conference on Theory of Computation.*

Cheeseman, P., Kelly, J., Self, M., Stutz, J., Taylor, M. and Freeman, D. (1988). AUTOCLASS: A Bayesian Classification System, In *Proceedings of the Fifth International Machine Learning Conference.*

Dao, S., Zhang, Y., Shek, E., and van Buer, D. (1999). High-speed Digital Wireless Battlefield Network, In *Proceedings of SPIE - Digitization of the Battlespace IV*, 3709:129-137.

Dao, S. and Perry, B., (1996). Information Dissemination in Hybrid Satellite/Terrestrial Networks, IEEE *Data Engineering Bulletin*, 19(3):12-18.

Fisher, D. (1987). Knowledge Acquisition via Incremental Conceptual Clustering", *Machine Learning*, 2:139-172.

Fox A. and Brewer, E., Reducing WWW Latency and Bandwidth Requirements via Real-Time Distillation, In *Proceedings of the Fifth International World Wide Web Conference.*

Fox, A., Gribble, S., Brewer, E., and Amir, E. (1996). Adapting to Network and Client Variability via On-demand Dynamic Distillation, In *Proceedings of the Seventh International Conference on Architecture Support for Programming Languages and Operating Systems.*

Floyd, S., Jacobson, V., Liu, C., McCanne, S., and Zhang, L. (1995) A Reliable Multicast Framework for Lightweight Session and Application Layer Framing, In *Proceedings of ACM SIGCOMM.*

Katz, R., Brewer, E., Amir, E., Balakrishnan, H., Fox, A., Gribble, S., Hodes, T., Jiang, D., Nguyen, G., Padmanabhan, V., and Stemm, M. (1996). The Bay Area Research Wireless Access Network (BARWAN). In *Proceedings of Spring COMPCON Conference.*

Macker, J., Klinker, J., and Corson, M. (1996). Reliable Multicast Data Delivery for Military Networking, In *Proceedings of IEEE MILCOM '96 Conference.*

Macker, J., and Adamson, R. (1999). The Multicast Dissemination Protocol Toolkit, In *Proceedings of IEEE MILCOM '99 Conference.*

Michalski, R. and Stepp, R. (1983). Automated Construction of Classifications: Conceptual Clustering versus Numerical Taxonomy, *IEEE Transaction on Pattern Analysis and Machine Intelligence*, 5:219-243.

Shek, E., Dao, S., Zhang, Y., van Buer, D., and Giuffrida, G. (2000). Intelligent Information Dissemination Services in Hybrid Satellite-Wireless Networks, *ACM Mobile Networks and Applications (MONET) Journal*, 5(4):273-284.

Shekhar, S., Liu, D., and Fetterer, A. (1996). Genesis: An Approach to Data Dissemination in Advanced Traveler Information Systems, *Bulletin of the Technical Committee on Data Engineering: Special Issue on Data Dissemination*, 19(3).

Weiss, M. (1996). *Algorithms, Data Structures, and Problem Solving in C++*, Addison Wesley.

Zhang, Y. and Dao, S. (1996). Integrating Direct Broadcast Satellite with Wireless Local Access, In *Proceedings of First International Workshop on Satellite-Based Information Services.*

# Index